WRITING AND LEARNING IN THE SCIENCE CLASSROOM

Science & Technology Education Library

VOLUME 23

SCOPE

The book series *Science & Technology Education Library* provides a publication forum for scholarship in science and technology education. It aims to publish innovative books which are at the forefront of the field. Monographs as well as collections of papers will be published.

The titles published in this series are listed at the end of this volume.

Writing and Learning in the Science Classroom

by

CAROLYN S. WALLACE
The University of Georgia,
Athens, GA, U.S.A.

BRIAN HAND
Iowa State University,
Ames, IA, U.S.A.

and

VAUGHAN PRAIN
LaTrobe University,
Bendigo, Victoria, Australia

KLUWER ACADEMIC PUBLISHERS
DORDRECHT / BOSTON / LONDON

A C.I.P. Ctalogue record for this book is available from the Library of Congress.

ISBN 1-4020-2017-1 (HB)
ISBN 1-4020-2018-X (e-book)

Published by Kluwer Academic Publishers,
P.O. Box 17, 3300 AA Dordrecht, The Netherlands.

Sold and distributed in North, Central and South America
by Kluwer Academic Publishers,
101 Philip Drive, Norwell, MA 02061, U.S.A.

In all other countries, sold and distributed
by Kluwer Academic Publishers,
P.O. Box 322, 3300 AH Dordrecht, The Netherlands.

Printed on acid-free paper

Printed in the Netherlands.

Table of Contents

CHAPTER 1

INTRODUCTION: DOES WRITING PROMOTE LEARNING IN SCIENCE?

Carolyn S. Wallace, Brian Hand, and Vaughan Prain

Most of us have had the experience of thinking about concepts in new ways or discovering gaps in our knowledge through the act of writing. The idea that learners could generate new science knowledge through composing text was proposed in the early 1970s. Science writing involves the use of reasoning skills to organize information, describe scientific phenomena, create knowledge claims, and formulate an argument. Thus, writing has potential for fostering content learning. Yet, research evidence on the specific character of science knowledge learned through writing, and the mechanisms through which the learning takes place has remained somewhat elusive over the past three decades. Writing in 1984, Applebee asserted, "At one level, most authors begin with the assumption that writing about a topic helps us to understand that topic better. At another level, we know almost nothing about the nature of the understanding that develops . . ." (p. 590). A decade later in 1994 (p.885), Holliday, Yore, and Alvermann noted, "Currently, conceptions of writing to learn appear to be embryonic and fragmented."

The authors of this volume formed a research team in the mid nineteen nineties to investigate writing to learn in science classrooms. Since that time, we have been fascinated by the question of whether and how the act of writing promotes science understanding. In this volume, we bring together the accumulated wisdom of many scholars in the writing to learn field and results of our own research program to synthesize implications and meaning for teachers and students of science. Our research on writing to learn science has been conducted within the messy context of science classrooms. Therefore, we view this book as a synthesis of research that can directly inform classroom practice.

WRITING AS A MEANS OF DOING SCIENCE

This book will argue and present evidence that writing does indeed foster science learning, although many contextual factors in the science environment influence the nature and degree of that learning. Writing is one mode of "doing science," just as hands-on laboratory work, Internet research, reading, or oral discourse constitute other ways of doing science. Writing is an essential activity that all students of science need to engage in to completely focus their scientific understandings. The idea that writing is a core science activity stems from theories of the relationship between language and learning (Britton, 1970; Gee, 1990; Halliday, 1993; Halliday & Martin, 1993; Norris & Phillips, 2001). Recent socio-cultural constructivist perspectives on learning emphasize the importance of interpretation and meaning for learning from experience. Halliday (1993, p. 93) asserts that, "The distinctive characteristic of human learning is that it is a process of making meaning- a semiotic process." This semiotic process is shaped, conceptualized, transformed, and reproduced through language. Thus, language takes a central role in the semiotic process, and creates learning. As Halliday (1993, p. 94) further explains, "Language is not a domain of human knowledge (except in the special context of linguistics, where it becomes an object of scientific study); language is the essential condition of knowing, the process by which experience *becomes* knowledge."

Writing and reading are critical elements of both the professional scientific enterprise and school science. Texts are necessary to communicate what has been learned before and what is currently being generated as new or reconstructed knowledge. Norris and Phillips (2001) posit that text is essential for the production and existence of scientific theory. Any attempt to express, reconstruct, or understand theory without the support of text quickly exceeds the limits of human memory, expressive ability, and attention. Not only verbal text, but recording in mathematical symbols or graphs, has evolved as method to condense complex scientific information into other forms, and is therefore a necessary practice (Halliday & Martin, 1993). Scientific thoughts are simply too complex to have meaning outside the existence of text. Norris and Phillips (2001, p. 24) suggest, "a conception of scientific literacy is needed that encourages a focus on mastering the many literate modes of expression required to portray and to acquire scientific knowledge."

In a similar vein, classroom learning in science cannot proceed without enculturation into scientific texts. Because science is a constructed enterprise, individual students cannot derive scientific explanations or theories solely from their own experience (Driver, Asoko, Leach, & Mortimer, 1996; Millar, 1998). It would be impossible for learners to recapitulate hundreds of years of experimentation and discourse that has resulted in powerful contemporary scientific ideas. Further, it would be a disservice not to provide learners with access to the explanations, elaborations, and discoveries of others. A further point is that literacy is essential for classroom communication. Text, as well as verbalization, is an important component of the social construction of scientific knowledge (National Research Council, NRC, 1996).

Science, Scientific Literacy and Using Language in Science

As researchers in the uses of language in the science classroom, we adhere to many of the values for science education specified in the National Science Education Standards (NSES, NRC, 1996). First, we consider science to be a way of thinking and a set of intellectual values, as well as, a collection of concepts, theories, and principles socially constructed over time. As stated by the NSES (p. 21), "science is a way of knowing that is characterized by empirical criteria, logical arguments, and skeptical review." Second, we believe that science is for all students, not just those who are scientifically talented. Students at all stages of development should have opportunities for rich experiences in science, including various forms of scientific writing. There should be high expectations for all students in terms of written communication and clear expression of understandings.

Third, we view scientific literacy as encompassing both knowledge of scientific concepts and a deep understanding of how scientific investigation relates to knowledge production in science (the nature of science). Recent discussions of what constitutes a scientifically literate person are diverse. Some examples of scientific literacy that are often cited in the literature include a knowledge of science vocabulary, using scientific information effectively in the real world, or being able to read and write scientific texts (Norris & Phillips, 2001). Our view is that one essential element of scientific literacy is to comprehend the evaluativist stance of science; that is, to have sufficient metacognitive skills to examine data in the context of a question, to make a knowledge claim related to the data, and to provide evidence for that knowledge claim in light of theory or explanation. The ability to perceive sensory information, compare it to prior knowledge, and generate meaning is fundamental to the process of knowledge construction in science (Osborne & Wittrock, 1983). Thus, an important element of scientific literacy is the ability to understand the relationships among questions, data, claims, and evidence, not only to understand the work of scientists, but also to learn how to learn science, to have the capacity to generate new scientific meanings for oneself.

Fourth, we believe that students can and should utilize a wide range of discourse types in scientific communication. Since the writing to learn movement took hold, there has been a debate in the literature as to whether students should use informal language to express their science understandings or learn to write in traditional scientific genres, like experiment and report (Berkenkotter & Huckin, 1995; Halliday & Martin, 1993; see also Chapter 4, this volume). Students need opportunities to use multiple discourse forms in scientific language use. Everyday language might be used in a peer group discussion or in informal journal writing to generate early understandings of scientific phenomenon. However, scientific language is ultimately more powerful for generating robust understandings than everyday language (Gee, 2001). Hybridized language or that which bridges everyday language and scientific language can be taught to help students unpack scientific language and gain an understanding of its purpose.

The Science Writing Heuristic (see Chapter 6), for example, scaffolds students to represent their own science understandings as more canonical forms including questions, claims and evidence. Much of our research has centered on teaching students to translate their science understandings into other forms of language for different audiences (see Chapters 5, 6, and 7, this volume). The translation of science ideas into multiple discourse forms provides students with opportunities to become flexible and fluid in their scientific knowledge. It is sometimes beneficial for students to communicate in traditional scientific genres, such as report or biography.

Writing and the Construction of Meaning

The paradigm shift from behaviorism to social constructivism as a philosophical, psychological, and social theory for learning illuminated the possibilities for writing as a means of learning science. The preeminence of a theory of science learning as a generative process (Osborne & Wittrock, 1983) paralleled both a view of writing as a constructive problem-solving endeavor (Bereiter & Scardamalia, 1987), and a view of science writing as a process of creating science knowledge, rather than recounting knowledge already well understood (Bazerman, 1985; Berkenkotter & Huckin, 1995; Halliday & Martin, 1993). It is now widely accepted that learners construct their understandings of both science and writing through the social negotiation of meaning. It is precisely this emphasis on meaning generation as a semiotic process that holds the key to learning science through writing.

As Vygotsky (1978) posits, learning may be defined as giving verbal shape to sensory impressions. As humans, we come to understand through the process of verbalization. Learning involves manipulating verbal ideas, patterns, and images. Language is the medium used to represent these images. By talking to ourselves and others, ideas take on reality. The composing process is an extension of the natural inclination to generate verbal meaning. It is impossible to compose without generating an understanding of what one is saying. In science, we receive sensory impressions of scientific phenomena. We come to understand these phenomena as we connect new information to that already in memory and generate new meanings (Osborne & Wittrock, 1983). Verbalization causes us to generate meaning in the terms of a language of which the aim is to communicate to others. Thus, there is an extremely close connection between science learning, verbalization, and writing.

Britton (1970) first posited the idea that speech, thinking, and writing are highly interconnected. He asserted that the writing of young students, or early drafts of more experienced writers are closely related to speech. He termed these informal writing products *expressive* to convey the idea that they express understandings that are incompletely formed in the writer's mind. The construct of expressive writing led the way for developing the writing-across-the-curriculum movement of the late 1970s and early 1980's. Language scholars and subject area teachers came to value the processes of informal writing as a way for students to manipulate, explore, and question subject matter content. The writing across the curriculum movement established many of the salient writing practices in our math, science, and social studies classrooms today including journals, freewriting, essays, creative writing, question writing, cartoons, and explanations.

Another pioneering language scholar, Janet Emig (1977) viewed writing as a unique form of learning, unlike reading or listening. She points out that writing involves three distinct modes of cognitive interaction: (a) the enactive form, learning by doing; (b) the iconic form, generating an image; and (c) the symbolic form, representing ideas as verbal symbols. Writing requires a coordination of eye, brain, and hand activity, and utilizes both the left and right areas of the brain. Emig asserts that the personal commitment involved in writing makes it especially engaging. Further, writing must be detailed and compact, so it causes a reformulation of ideas. Finally, Emig notes that because writing takes a permanent form on the paper, it allows the writer to blend learning processes and products together.

As compelling as these early theoretical arguments have been, research on the efficacy of writing to produce science learning has been inconclusive until very recently. Rivard's (1994) review article on writing to learn in science pointed out some of the conceptual and technical barriers to conducting rigorous research in this area. Links between writing and specific types of learning have not been not well understood. For example, does writing produce conceptual change, more sophisticated reasoning or both?

He also cited a need for more studies in which the participants were trained in metacognition. Without an understanding of how writing might assist learning, participants may not have been able to take advantage of the writing to learn interventions implemented in the research. In addition, subject-task-topic interactions require more research. Although early research indicated that learning from writing was very sensitive to the specified task (Langer & Applebee, 1987), few of these types of interactions had been explored.

In the last few years, new syntheses of hypotheses for learning through writing (Galbraith, 1998; Klein, 1999), as well as fresh approaches to researching science learning through writing (for example, Mason & Buscolo, 2000), have contributed significantly to the field. The purposes of this volume are to: (1) review the contemporary research of others, within the context of an early research backdrop; (2) explore recent theories, hypotheses, and models for learning science from writing; (3) create a rationale for the importance of writing as a form of science literacy; (4) synthesize our own contributions to the research; and (5) apply research results and evidence directly to implications for classroom practice, teacher education, and future research in writing to learn. We hope that this volume will be used by both researchers and practitioners of science education, as we believe there is a fine line between classroom inquiry on writing to learn and good classroom practice.

OVERVIEW OF THE CHAPTERS

Chapter 2, by Carolyn S. Wallace (formerly Carolyn W. Keys), "Evidence from the Literature for Writing as Mode of Learning Science," will provide the reader with background knowledge on research in the field. The chapter will briefly outline the roots of writing to learn in science from the writing across the curriculum movement. It will then review studies providing the most contemporary evidence for learning through writing. My focus will be to highlight salient research conclusions that build a knowledge base from which other studies may follow. The chapter will also elucidate research studies of how learners are enculturated into the practice of written scientific discourse. Relevant studies on the relationship between teaching practices and learning will be discussed.

Chapter 3, "Cognitive, Constructivist Mechanisms for Learning Science through Writing," will outline the basic psychological functions for the possibility of learning through writing. It will explore the question: What cognitive mechanisms can be posed to explain how children learn science from writing activities? The work of leading cognitive psychologists working in this area will be explored, including Emig, Flower and Hayes,

Langer and Applebee, Scardamalia and Bereiter, Galbraith, and Klein. The thinking of these scholars presented in chronological fashion indicates how the theory of writing to learn has been expanded over the past few decades. In this chapter, the author, Brian Hand, will explore a synthesis of these hypotheses as a best explanation of how learning science could be accomplished through writing.

In Chapter 4, "The Role of Language in Science Learning and Literacy," will explore the ways in which language is important to science learning. The chapter will outline the debate among science language scholars as to what types of writing are most appropriate in school science. The author, Vaughan Prain, will present the argument that everyday language is an important component of scientific literacy and that students need opportunities to express their understandings in diverse forms. The role of multi-modal forms and technology in creating science learning will also be discussed.

Chapter 5, "A Research Program on Writing for Learning in Science, 1992-2002," will include a description of the various types of writing practices Hand and Prain have been implementing in collaboration with classroom teachers in both Australia and the United States. The chapter reports on two distinct research phases, an initial qualitative phase in which the researchers and teachers developed various writing to learn strategies and a second quantitative phase, in which the effectiveness of specific writing treatments were tested. The effects of instructional tasks, such as planning for writing are evaluated.

Chapter 6, "Using a Science Writing Heuristic to Promote Conceptual and Metacognitive Understandings," will report on the creation and use of a heuristic tool to promote understanding from laboratory. This heuristic tool, the Science Writing Heuristic (SWH) provides both teacher and student activities to foster meaning negotiation through speech and writing. The chapter includes a discussion of the theoretical underpinnings of the SWH, as well, as a synthesis of research studies on its efficacy in the science classroom. Research in several classroom settings conducted by Wallace and Hand will be synthesized.

Chapter 7, "Children's Views of Writing to Learn in Science," describes key attributes of writing to learn in science, according to students from grades 7 to 12. Based on extensive interviews with children in several research projects, Wallace and Hand depict the tasks and cognitive processes children find most useful in writing to learn projects. Implications for the classroom are discussed.

Chapter 8, "Teachers' Perceptions of Writing to Learn Strategies," summarizes Hand and Prain's research on how teachers have adapted writing to learn for their own classroom contexts. Two in-depth, qualitative studies reflect the benefits and challenges of implementing writing for learning in the science classroom. Several assertions are presented that represent teachers views of their own changing roles, their concerns about student learning, and the creation of their own strategies for developing writing to learn activities and assessments. The delicate balance between guiding students and promoting independent student thinking is discussed.

Chapter 9, "Breakthroughs, Classroom Implications, On-going, and Future Research," summarizes what we have determined to be the major contributions of our research in the field. This final chapter emphasizes what is most important to students and teachers about writing to learn and synthesizes our research findings into science classroom applications. In this chapter we describe our current projects and new questions arising from our work. Limitations of our research and questions for future research are included to suggest new directions for research in writing to learn in science.

CHAPTER 2

EVIDENCE FROM THE LITERATURE FOR WRITING AS A MODE OF SCIENCE LEARNING

Carolyn S. Wallace

Most scholars cite the early work of Britton (Britton, 1970; Martin, 1992) and Emig (1977) as the beginning of the writing to learn paradigm. These researchers first posited that writing causes one to concretize knowledge into verbal symbols, and thus, new subject matter could actually be learned while writing. This idea became very popular in universities throughout the 1980s, resulting in a "writing across the curriculum" movement, where writing was incorporated into all types of subject matter courses. Concurrently, Britton's notion of expressive or informal writing became popular for promoting writing for understanding in formats such as journals, mini-essays, questions, and freewrites (Martin, 1992). Eventually, the writing across the curriculum movement migrated down into some secondary and elementary school science classrooms (Rivard, 1994). Much of the early research on writing to learn took the form of action research by college professors, who did small scale quantitative and qualitative studies to determine if writing improved disciplinary learning. The results were often inconclusive due to small sample sizes and other technical problems with the studies. However, almost universally, student learners expressed that they had found value in writing to learn.

A study of writing in historical geology provides an example of early studies of writing to learn science (MacDonald & Purdy, 1989). MacDonald and Purdy's assignment included observation and description of a local geological feature, an explanation of the possible history of the feature, and a peer critique in preparation for revising to a final draft. The authors' qualitative and anecdotal evidence of the efficacy of the writing assignment indicated that students used keen observation skills and newly acquired geology knowledge to write rich descriptions of their chosen feature. However, only about one-fourth of the papers included interpretive speculations or hypotheses about the origin of their features. The majority of the students in the class (74%) rated the assignment highly and many gave specific comments about the benefits of the writing assignment, "As a whole the assignment was invaluable. I felt the exercise was the most effective learning experience in the course," (MacDonald & Purdy, 1989, p. 119).

The MacDonald and Purdy (1989) study provides a good example of the issues raised by early writing to learn research. First, the vast majority of students who participate in writing to learn find it to be a valuable learning experience. Thus, writing has become known as a "promising" learning activity (Rowell, 1997). Second, students did not naturally include interpretations as part of their writing, pointing to the idea that to realize its full potential, writing to learn must be explicitly taught. Third, this study like many others, hints at the complexity of the mechanisms that might be involved in learning science from writing. For example, the authors noted a wide variety in the quality of the written reports, suggesting that students used a variety of cognitive and metacognitive strategies in approaching this assignment.

Issues such as these fueled the next wave of research on writing to learn science that was aimed at more specific questions: (1) Does writing really increase science learning and, if so, what types of learning might it improve? (2) What might be the cognitive and metacognitive mechanisms involved in this learning? (3) How do learners become enculturated into writing in scientific genres? (4) What classroom activities and/or tasks promote learning from writing in science and why do they? Several comprehensive reviews of research on writing to learn in general and writing to learn in science have addressed some aspects of these four questions, although they all suggest that further research is needed (Applebee, 1984; Berkenkotter & Huckin, 1995; Klein, 1999a; Rivard, 1994; Rowell, 1997; Tynjala, Mason, & Lonka, 2001). In the discussion that follows, I review the most recent research available on these topics as a background for the reporting of our own research studies that follow in subsequent chapters.

THE IMPACT OF WRITING ON CONCEPTUAL UNDERSTANDING

Researchers have long been interested in whether writing increases science conceptual understanding or promotes conceptual change. Evidence for the impact of writing on conceptual knowledge has most often been investigated through pre and post test designs, in which the tests typically include conceptual multiple choice questions, writing short essays, or writing explanations. Several recent studies provide evidence that writing does promote science conceptual understanding, although this is not universally the case. In one of the most positive studies, Mason and Boscolo (2000) investigated the learning of 36 fourth grade students in an experimental writing to learn class and 20 control students in a separate class. The control students did all the same activities as the writing students, except journal writing in a unit on photosynthesis. The researchers found that the experimental group performed significantly better than the control group on three measures of conceptual understanding, open-ended questions, transfer questions, and short text (p always < .01).

In a follow up study, Boscolo and Mason (2001) investigated whether and how fifth grade students trained in writing to learn for history could transfer their metacognitive and attitudinal understandings of the benefits of writing to science. The researchers found that students in the experimental writing group did use writing in the science context (human circulatory system) to give explanations, elaborate their ideas, and explore gaps in their own knowledge. The experimental group outperformed the control group on conceptual change measures including, topic knowledge, open-ended questions and transfer questions, but there was no difference on short text writings. The study offers support for the usefulness of writing to promote science conceptual change when children understand its benefits. Similarly, Tucknott (1999), working with fourth grade students, found a 51% gain on high-level comprehension test questions, using a simple pre/post design for simple machine concepts.

In a qualitative study, Fellows (1994) investigated the learning of sixth grade students involved in a conceptual change unit on matter. She traced improvements in students written explanations over the unit. These explanations gradually became closer to the targeted scientific explanations. Fellows claimed that more opportunities for writing explanations led to better logical arguments and conceptual change. During the lesson cluster when students wrote more explanations, more conceptual growth was noted. Keys (1994) found that knowledge bases for physical science concepts improved for five of six target students when collaborative pair writing was implemented in the classroom.

Two recent experimental studies indicate more qualified support for writing to promote conceptual change. Rivard and Straw (2000) randomly assigned eighth grade students to four groups to investigate the influences of peer group discussion and writing on ecology concepts. They found that the talk and writing group significantly outperformed all other groups on delayed post tests of simple and integrated knowledge. The writing only group did not perform significantly better than the control group on either measure. The talk only group outperformed the control group on simple knowledge, but not integrated knowledge. The authors, examining these results and their qualitative data, concluded that talk was most important for distributing knowledge, while writing was important for manipulating, consolidating, and integrating knowledge.

Also using a quantitative design, Klein (2000) investigated the explanations of 36 fourth, sixth, and eighth grade students following experiments on buoyancy and balancing. He found that after journal writing, the explanations of 10 of the 36 students for buoyancy and 5 of the 36 students for balancing improved. Thus, writing improved learning significantly for only a minority of the students. Klein's further investigation of the mechanisms of that improvement will be discussed below. In a similar study with pre-service elementary teachers, Klein (1999b) determined that 43% of the students studying buoyancy and 13% of the students studying forces improved the complexity of their explanations through additional journal style writing. Thus, Klein offers modest support for the efficacy of writing in improving conceptual understanding.

In summary, the findings suggest that writing has the potential to increase conceptual understanding, but not in all cases does writing lead to greater conceptual understanding. The results of these studies are suggestive that when writers, either intuitively or through instruction, recognized the value of writing for improving conceptual understanding, they undertook the cognitive tasks through writing that would lead to increased understanding. For example, the children in Mason and Buscolo's (2000) study were explicitly taught the value of reflective journal writing, and they performed very well on tests of conceptual understanding. In Klein's (2000) study, only a small portion of the children increased understanding through writing. Perhaps the other children did not increase their understanding, because they were not aware of the cognitive strategies for doing so.

Evidence for Cognitive and Metacognitive Mechanisms

A companion question to whether writing improves conceptual understanding is how this learning might occur. Models and hypotheses for how writing fosters learning appears in Chapter 3 of this volume. Some empirical literature to date provides insights into the cognitive and metacognitive processes of learning through writing.

One of the most salient features of writing that may promote conceptual understanding is its potential to generate metacognitive thought. When learners write, they discover what it is they think, so that they come to a better understanding of what they know and what remains a gap in their knowledge. This metacognitive awareness may then stimulate other cognitive strategies, such as searching for new knowledge, or reviewing text. Thus, metacognition is a form of learning that is produced by writing, while at the same time, it is a catalyst and, thus, part of the process for content learning. Research studies that show conceptual growth through writing often have indicated parallel metacognitive growth (Mason & Buscolo, 2000). While few direct connections can be made (Klein, 1999), an association between increased metacognition and increased conceptual learning suggests that metacognition plays a role in the process.

A research study by McCrindle and Christensen (1995) provides strong support for the role of metacognition in learning from writing in biology.

College freshman were divided into control and treatment groups during five weeks of biology lab. Students in the control group wrote traditional laboratory reports about their lab activities, while students in the treatment group wrote learning journal entries. The treatment group received direct instruction on the use of learning journals for reflecting both on their cognitive knowledge about the labs, or what they had learned, and their metacognitive knowledge, or how they had come to learn it.

Results indicated a strong advantage for the learning journal group on several measures. The participants in the learning journal group had more sophisticated conceptions of the nature of science learning, including understanding and interpretation, than the control group. Interviews revealed that the learning journal group used more metacognitive strategies, such as checking for understanding. On a cognitive strategy task, the experimental group used more elaboration and organizational strategies and fewer rehearsal strategies than the control group. The experimental group also performed significantly better on a task designed to elicit complex and abstract knowledge. Finally, the experimental group outperformed the control group on the course final exam with an effect size of 0.79 of a standard deviation. This study implies that a writing task specifically designed to generate metacognitive thought promotes not only metacognition, but also more sophisticated cognitive strategies and better learning outcomes. Although the control group engaged in writing, traditional laboratory reports tend to obscure, rather than elicit the processes involved in scientific thinking (Keys, 2000).

Mason and Buscolo (2000) investigated metacognitive awareness for fourth grade students in control and writing treatment groups, for which conceptual gains are described above. The writing group achieved significantly higher scores on a measure of metaconceptual awareness and demonstrated higher agreement that writing can be used for reflection on what is and is not understood. The authors concluded that representation of the value of the writing changed, just as the representation of the biological concepts. Similarly, Buscolo and Mason (2001) determined that students with the opportunity to write about their conceptions and monitor developing understandings achieved higher levels of metaconceptual awareness, as well as, more sophisticated understandings about human circulation. They claim that writing allowed these students to be aware of how they integrated knowledge into their conceptual structures.

Because of the difficulty of directly investigating cognitive strategies for learning while writing, few studies have attempted to probe this "black box" (Klein, 1999). Klein (2001) investigated learning taking place through writing in association with buoyancy and balancing problems. The fourth, sixth, and eighth grade students in his study each carried out a science experiment, stated their explanations of the phenomena, then wrote journal style notes while thinking aloud. They then wrote a second explanation of the phenomena. The think aloud protocols and writing of students who improved their explanations was compared to those who did not improve their explanations. Several cognitive strategies including writing operations, transitions, text features and content generating strategies were explored. Factor analysis allowed an aggregation of data into the most striking differences between those who improved their explanations and those who did not. Although only a minority of students did learn from writing, as exhibited by better explanations, their learning actions provide clues into how such an improvement might take place.

Klein's (2000) data indicated that three clustered factors significantly predicted learning from writing, including Brainstorming, Search from Text, and Search from Experiment. Search from Text meant that these writers reviewed text that they had already produced in preparation for the generation of new text. Search from Experiment indicated the action of reviewing experimental results to generate new text. Thus, searching for additional information, as well as, reviewing what has been stated both were characteristics of those who learned from writing. Klein concludes that for elementary students, learning is a strategic, goal-directed phenomenon. The cognitive strategies that contribute to learning from writing are independent from one another and cumulative in their effects.

In summary, research studies to date show the high level of connectivity between cognitive and metacognitive strategies and increased learning from writing. There is evidence that metacognition, or monitoring one's own understanding and new knowledge integration, supports conceptual growth. The potential of writing for making these metacognitive connections clear to the learner is great. Cognitive strategies that may lead someone to learn from writing include purposeful and deliberate searches for new information, as well as syntheses and evaluation of what already has been written. It seems probable that to capitalize on learning from writing, learners will need to be taught how and when to use these cognitive and metacognitive strategies.

Enculturation into Science Writing Discourses

The corollary to "writing to learn science" is "learning to write science." How scientists use language to express their ideas or the linguistics of science is an entire field of study (Bazerman, 1985; Berkenkotter & Huckin, 1995; Halliday & Martin, 1993; Myers, 1990). The contemporary view is that written language is not merely an expression of science that is already known, but that writing actually creates science as it takes form on paper (Bazerman, 1985). There has been some recent research on how K-16 students learn scientific forms of writing. The questions have focused around how students develop, appropriate and use cultural scientific forms of expression, known as genres. Although not a major focus of this volume, the literature on enculturation into scientific writing is reviewed briefly here, since learning to write like a scientist may be considered an aspect of learning science. Our view is that enculturation into scientific writing is necessary and desirable, as long as students also have opportunities to express science ideas in informal modes, and to translate their ideas into multiple discourse forms. Varelas, Becker, Luster, & Wenzel (2002) have recently reported that young children demonstrate strong science understandings when these are linked with familiar, rather than scientific discourse genres.

Studies with primary grades students indicate children begin to appropriate scientific genres at a very early age (Kamberelis, 1999; Shepardson & Britsch, 2001; Varelas, Pappas, Berry, & Rife, 2002). Kamberelis investigated the genre development of fifty-four kindergarten, first, and second grade students. He found that although genre knowledge of a science report was less well developed than genre knowledge of a narrative, the children expressed knowledge of science genre characteristics, including observation, description, and using technical language. His findings indicate that young children absorb cultural norms for genre from the environment. He stated:

> Based on the findings from this study, however, it is clear that children are by no means genre somnambulists during the first few years of school. Like their knowledge of symbolic aspects of written language, their knowledge of genres is complex and multiplex (Kamberelis, 1999, p. 445).

Varelas, Pappas, Barry, and Rife (2002) working with first and second grade students similarly found that students could express their ideas about evaporation in the form of explanations. The researchers suggested that the use of words and drawings together were complimentary semiotic forms; those ideas that could not be expressed in words could be expressed in drawings. An investigation of children's science journal writing by Shepardson and Britsch (2001) indicated that very young children bring their own worldviews into scientific writing. First graders in their study drew on imaginary worlds and prior experience worlds, as well as investigative worlds to interpret the results of hands-on activities. For example, one child, Deanna, wrote a story about a duck walking on the beach in response to a science activity on dissolving. Some first graders, though, seemed to have appropriated characteristics of a science report, similar to the findings of Kamberelis (1999). By fourth grade, the students in Shepardson and Britsch's study had abandoned references to imaginary worlds in their journal writing, and drew only on prior experience and investigative worlds. In addition, their journal texts indicated influence of cultural norms for scientific genres. Most chose to display data in tables with columns and rows. They attempted to both explain and interpret their data. The authors conclude that familiarity with outside texts, such as those seen in science books, may have been employed in the journals.

Not many studies of science discourse appropriation have been conducted at the middle school level. Keys (1999) found that without explicit instruction in experimental report writing, middle school students tended to write more observation and description type clauses than inferential or explanatory clauses about authentic investigations in zoo animal behavior and stream studies. The close association of observation and interpretation, although made distinct by scientists (Kelly, Chen, & Prothero, 2000), is an important feature of scientific writing. Scientists tend to write the meanings for their findings in close physical and conceptual proximity to their presentations of data (Halliday & Martin, 1993). When the middle school students who had intuitively adopted this style wrote reports, they were rich in scientific detail and hypothesizing.

At the high school level, Audet, Hickman, & Dobrynina (1996) found that high level physics students studying fractals appropriated many sense-making forms of communication into their learning logs, including interpretation and story telling forms of writing. The learning logs were very effective communication tools used in the building of socio-linguistic norms in the science classroom. Through the learning logs, the teacher could monitor student thinking, as well as, motivate new questions. Thus, students with a high interest in the topic appear to easily appropriate informal forms of scientific writing, and capitalize on its value for constructing understanding.

The ease with which students appropriate informal science writing, does not seem to transfer into more formal report writing. Kelly and Chen (1999) investigated high school students' appropriation of the scientific cultural norms for supporting claims with evidence in reports on the physics of sound. In the unit explicit instruction for writing a technical report based upon scientists' recommendations, was provided to the students. The researchers report that the students revealed a partial engagement with the intended task. They attempted to follow the scientific genre presented to them, they included claims, which were sometimes supported by evidence, and they included supporting data. Although there was a wide variety of performance, students often copied the general genre style, but their arguments lacked substance. Some did not use much evidence to support their assertions. The authors suggest that a lack of congruence between classroom activities and scientific social practices rendered the instruction, and therefore the reports, incomplete. To us, it suggests that students need to unpack their understandings of scientific concepts through everyday language, before they can be expected to appropriate abbreviated and embedded forms of scientific communication, such as that found in a technical report.

Similarly, Campbell, Kaunda, Allie, Buffler, and Lubben (2000) found that entering university freshman had difficulty determining what to include in a scientific laboratory report about a physics investigation. Without explicit instruction in writing, students had vague and incomplete notions of appropriate writing for the report. They inadequately described background information relevant to the investigation, they gave little attention to describing specific laboratory procedures and data collection processes, and no justification for decisions made during the investigation. Interestingly, when students changed their procedures during the investigation due to group discussion, they tended to report their original procedures, rather than the ones they finally used. The data indicates that students are unfamiliar with and need enculturation into scientific report genres. The authors outline implications for college science writing instruction.

In contrast, Florence and Yore (2002) reported on the high efficacy of co-authoring as a practice for graduate and postgraduate interns to learn scientific writing. While writing together, scientists are able to enculturate novices into normed practices, such as, outlining, drafting, and revising. The authors state, "Supervisors provided just-in-time tutoring to students with real needs and in the context of real problems," (Florence & Yore, 2000, p. 23). The study shows the value of situated learning and co-authoring between novice and expert in learning to write science.

Taken as a whole, these studies seem to indicate that students at a very early age begin the incorporation of the ideas of scientific genre. While young students easily master the basic ideas, as more detailed and cognitively demanding distinctions of scientific writing are needed, students find the genre conventions more difficult to appropriate. Writing about authentic and personally meaningful research appears to facilitate the process. These studies, as well as some of our own, suggest the tension between the need to organize, clarify, and unpack scientific ideas in everyday language and the demands of communicating in the accepted scientific genres. This tension is explored more thoroughly in Chapter 4 of this volume. Thus, it may be difficult to accomplish the goals of writing to learn science and learning to write science simultaneously.

THE INFLUENCE OF TEACHING STRATEGIES AND TASKS

It is difficult to link the influence of particular writing teaching strategies and tasks with particular learning outcomes in the ill-structured environment of the classroom, although there have been many calls to pursue this type of research (Holliday, Yore, & Alvermann, 1994; Rivard, 1994). Much of the work on the effects of writing for particular audiences, writing in conventional and non-conventional genres, planning for writing, and writing on the same topic in different modes has been carried out by members of our research team, Hand and Prain, and will be discussed in depth in Chapter 5 of this volume. We know that learners are very sensitive to the type of task they are asked to perform and the types of science learning that might be expected (Langer & Applebee, 1987). This important area of research was started in the mid-1980s through the seminal research of Langer and Applebee (1987). The researchers studied the way that writing to learn was implemented in science and social studies secondary classrooms and how these various tasks led to content area learning. They found that analytic writing tasks, such as essay writing, tended to develop those specific content points that are focused on in the writing, leading to deep, rather than broad understanding. If content is already well understood, there is no significant impact of writing on learning.

In a study of science writing implemented in a university program, Chinn and Hilgers (2000) found that most professors' assignments introduced students to those genres used by practicing scientists, including, literature search, laboratory report, peer review, critique, and oral presentation. Writing assignments could be clustered into two major categories: writing to learn content, such as outlines and summaries; and writing to learn disciplinary genres, such as literature review. They further found that professors' teaching strategies with regard to writing could be located on a continuum with three key reference points, instructor as corrector, instructor as journal editor, and instructor as collaborating scientist. The majority of teachers fell into the instructor as corrector category. They viewed themselves as the sole prescriber and audience for science writing. They portrayed writing critique as finding errors to be corrected and both they and their students considered the goal to be the achievement of a high grade. They did not focus assignments either on learning content or learning scientific communication. In one case study of instructor as corrector, the teacher put a great deal of time into meeting with students on an individual basis. Despite this effort, most of his students failed to grasp the fundamental structural and rhetorical features of scientific writing. In a small number of the classes studied, Chinn and Hilgers (2000) found a focus on peer review, hence, the label, instructor as journal editor. These professors attempted to enculturate their students into the importance of draft review by placing students in the role of peer reviewer. Students turned in their critiques of peers' rough drafts as part of their assigned work. Students had mixed reactions to this writing emphasis, although they found peer critique helpful, some students, especially women in the class, found it unnecessarily harsh and sometimes confusing.

The final approach found by the researchers was instructor as collaborating scientist. These professors emphasized writing as a multiple-purpose tool for learning and communicating science knowledge. The activities they assigned involved realistic or community-based topics and genuine audiences. Their assignments also involved collaboration. Students of these instructors indicated satisfaction and learning through the writing assignments. The study points to the clear benefits of reform-based approaches to science writing. Kelly, Chen, & Prothero (2000) also investigated the use of scientists' teaching strategies to enculturate college students into the norms of technical writing. They found that the scientists emphasized the following features of scientific writing: (a) the importance of arguments based on data sets, including the practical purposes for those arguments, such as funding and politics; (b) that science is grounded in fact, but not absolute, and (c) the importance of learning what an observation is and how it is distinct from interpretation. Thus, scientists invoked characteristics of science as a social discipline when teaching students how to write. The authors did not report on student learning outcomes in this study.

In summary, there is scant research in the area of how tasks affect learning from writing. From the research that is available, we may conclude that student learning is very sensitive to the types of writing tasks they are asked to perform. When constructivist-oriented practices, such as writing to a real audience or writing about authentic problems are used in the science classroom, more student involvement and satisfaction occurs. Links among writing tasks and some specific science learning outcomes will be explored in this volume.

CHAPTER 3

COGNITIVE, CONSTRUCTIVIST MECHANISMS FOR LEARNING SCIENCE THROUGH WRITING

Brian Hand

Throughout the last 30 years there has been much attention to exploring the cognitive processes associated with using writing as a learning tool. An article by Emig (1977) titled "Writing as a Mode of Learning" began an exploration into what actually takes place as an individual writes and learns from the act. As she said, "writing represents a unique mode of learning- not merely valuable, not merely special, but unique" (p. 122). Now, as then, recognition is given to the fact that we still do not know exactly the process through which writing promotes subject matter learning. As Holliday, Yore and Alvermann (1994, p. 886) point out "the verification of a theoretical model of writing to learn science that fully reflects contemporary views of science learning and writing is needed." It should be noted that while we do not yet have a current theory, we do have a richer set of possible explanations.

This chapter will outline the basic psychological foundations for the possibility of learning science through writing. The chapter is organized in the following manner. First, the interactive-constructivist position of learning theory that guides our ideas of writing as a learning strategy is presented. Second an overview of development of current understandings will be examined through an historical perspective starting with Emig (1977) and ending with Klein (1999). Third, a description of cognitive processes that frame our research questions is outlined.

THE INTERACTIVE-CONSTRUCTIVIST VIEW OF LEARNING

While most contemporary scholars ascribe to a constructivist view of learning in general, there has been much debate on the particular type of constructivist activities involved in learning. The various positions of the radical constructivists (von Glasersfeld, 1988), that knowledge is constructed at the individual level, and the social constructivists (Airasian &

Walsh, 1997), that knowledge is constructed at the community level, have been discussed in the literature. However, the interactive-constructivist position which incorporates elements of both radical and social constructivism, is that knowledge should be viewed as being personally and socially constructed (Driver, 1990) resulting in two forms of knowledge, one private and the other public.

Each individual has experiences with the world and other people and as a consequence is able to construct knowledge based on interactions within their social groups (Driver, 1986). That is the public aspect of knowledge construction. Meaning is made when an individual reflects on and makes sense of these interactions, which is the private aspect of knowledge construction (Henriques, 1997). Learning involves negotiating meaning of the knowledge both within the public forum and within the individual's own conceptual framework. The contrast between the various positions of constructivism has been described by Yore (2001). Yore locates four faces of constructivism, in terms of their various stances on an ontological worldview, an epistemic view of science, criteria for judging the knowledge constructed, the psychological locus of mental activity, the pedagogical locus of structure/control, and the linguistic discourse. Importantly, the interactive-constructivist position adopts the evaluativist view of science and includes nature as one constraint on knowledge. The evaluativist view posits that while science is a creative endeavor, it must hold up to the scrutiny of public critique and the nature of the physical world. The interactive-constructivist position locates the view of science firmly within the modernist position with all the attendant needs for claims, warrants and evidence. Thus, we believe that while diverse modes of discourse are important for understanding science, scientific canons of claims, warrants, and evidence form the core of scientific understanding. These constructions should be taught as part of the nature of science through writing to learn strategies.

HISTORICAL DEVELOPMENT OF CURRENT MODELS OF WRITING AS A LEARNING TOOL

Janet Emig's published an article in 1977 that signaled the beginning of a close examination of writing as a learning tool. This led to various models of the cognitive process of writing, including those of Flower and Hayes, Langer and Applebee, and Bereiter and Scardamalia in the 80s, and Klein and Galbraith in the 90s. Emig (1977) began her conversation by asserting that writing is not talk recorded, that is, writing serves a different function from talking. While pointing out a number of differences she indicated that "writing tends to be a more responsible and committed act," "with writing the audience is usually absent" and "because writing is often our representation of the world made visible, embodying both process and product, writing is more readily a form and source of learning than talking" (p.124). Borrowing from the ideas of Dewey, Piaget and Bruner, writing was viewed as a process that involved three categories of cognitive activity. These involved enactive learning, that is, learning by doing; iconic learning, that is, by depiction as images; and symbolic learning, that is, by restatement in words. Importantly, the act of writing was viewed as involving all three of those processes almost simultaneously and, thus, can be thought of as a multi-representational mode of learning.

A critical component of this multi-representational approach is that the act of writing allows for immediate feedback to be received "because the information from the *process* is immediately and visibly available as that portion of the *product* already written" (p.125). This constant form of feedback allows for, and in fact requires, the development of systematic connections and relationships between ideas expressed in the writing. This process, which is viewed as being much slower than talk, requires analysis and synthesis to occur. Analysis involves breaking down of "entities into their constituent parts," while synthesis involves "combining or fusing these, often into fresh arrangements or amalgams" (p.127). The overall impact of analysis and synthesis is a greater understanding as a consequence of having written the text. In summary, Emig's paper was a first effort to make the case for writing as a learning tool, as she delineated cognitive relationships between composing and conceptual knowledge.

During the 1980s, new models were developed that depicted writing as a problem-solving activity. Flower and Hayes (1980) introduced the concept of writing as problem-solving to explain how writing might serve learning. For them, the cognitive process theory was based on four points:

1. The process of writing is best understood as a set of distinctive thinking processes that writers orchestrate or organize during the act of composing.
2. These processes have a hierarchical, highly embedded organization in which any given process can be embedded within any other.
3. The act of composing itself is a goal-directed thinking process, guided by the writer's own growing network of goals.
4. Writers create their own goals in two key ways: by generating both high-level goals and supporting sub-goals which embody the writer's developing sense of purpose, and then, at times, by changing major goals or even establishing entirely new ones based on what has been learned in the act of writing (p.366).

By major or high-level goals, Flower and Hayes meant the larger themes of the text, whereas, sub-goals referred to local purposes or ways of tackling each section of the text. The acceptance of these four key points allowed a model of the writing process to be generated which moved beyond a linear conception or writing as a series of stages. Flower and Hayes believed that writing could be analyzed as a set of mental processes that are hierarchical, changeable, and embedded in nature.

The model proposed consisted of three major elements: the task environment, the writer's long-term memory, and the writing process (Flower & Hayes, 1980). The task environment is viewed as something outside of the individual, and includes the rhetorical problem or assignment and the growing text being produced. The writer's long-term memory is where knowledge of the topic, audience and writing plans are stored. The writing process deals with the three basic processes used within the act of writing. These are *planning, translating* and *reviewing*, which are all under the control of the writer. In proposing the model, recognition was given to the concept that there were two competing processes operating when writing was being undertaken. These were the use of knowledge of the content stored in long-term memory and the plan for dealing with the rhetorical problem. The act of writing was viewed as a balance between these processes and the actual text being generated.

Cognitive activity surrounding the subject matter content relates primarily to the third element of the model, the writing process. The writing process components include planning, translating and revising. These three processes form the central core of the act of writing. Each of these processes is made up of particular sub-processes. Planning involves three actions: (a) generating ideas, including retrieval from long-term memory; (b) organizing, which assists the writer in making meaning of the ideas and evaluating the affect of the rhetorical demands on the writing; and (c) goal setting, which involves the writer creating procedural and substantive goals to achieve the writing task. Translating involves representing the information generated in the planning phase in text. However, translation is viewed as being more than just a consideration of the rhetorical elements. It also involves global goals of transforming content into linear expression. Movement between these various phases of the process is controlled by a monitor, that is, the writer constantly monitors the amount of activity required in each of the phases.

Flower and Hayes viewed writing as a goal-directed activity involving the need to deal both with the process of writing and the content being addressed. This goal-directed activity is more than the broad goal of the overall task, but includes many sub-goals that assist in achieving the task. In essence, a network of goals is built up that allows the writer to complete the particular task. As such there exists a "struggle for influence" (p. 380) between the writer's goals, knowledge and current text within the writing process. Importantly for Flower and Hayes was the concept that there was some form of cycling between the "top level goals" (p.382) required to complete the task, and the sub-goals which impact on the on-going production of text.

In summary, Flower and Hayes proposed a model that emphasized a goal-driven process with a hierarchical view of top down goals. In completing the task, the writer was balancing the goals of the task with the content knowledge to be addressed and the text that was being produced. The overall result implicated writing to act as a learning tool.

In a landmark study, Langer and Applebee (1987) discussed how they thought writing was able to shape learning. As with the work of the authors described above, Langer and Applebee agreed that little work had been done to that point to determine specifically how writing was able to assist learning. They attributed this to two factors. The first was that the rhetorical presentation had been focused on audience analysis, that is, how to write to an audience rather than on content, while the second was the assumption that writing would lead to learning without there being supportive evidence for this position.

To study the value of writing as a learning tool, Langer and Applebee implemented a large study of teachers and students to examine classroom applications of writing. For them writing was viewed as something which could "provide varied and effective ways for students to think about and reformulate new learning and to integrate new information with their previous knowledge and experience" (p. 19). While they did not propose a cognitive model they did provide evidence of cognitive processes being used in completing the writing task. An interesting and important highlight of their study was that for classroom applications of writing they believed that the teacher's understanding of teaching and learning was a major determinant, that is, the teacher's conception of learning impacted greatly on how writing was used. Given this constraint they found that writing was used to serve three major functions (p. 41). These were:

1. To draw on relevant knowledge and experience in preparation for new activities
2. To consolidate and review new information and experiences
3. To reformulate and extend knowledge

The think aloud protocols used in Langer and Applebee's study showed that students were implementing a wide range of cognitive processes to complete writing tasks. These included hypothesizing, questioning, making metacomments, using schema both content and personal, and making global versus local comments. Importantly, they argued that not all forms of writing would promote the use of all these cognitive processes that they had identified. Writing activities that required the reformulation of content required much more thinking than review activities. The review activities required the writer to focus on specific content, while the reformulation activities required a more global focus and higher order thinking. Writing activities such as note-taking or answering study questions were viewed as review activities, while essay writing was seen to engage more reformulation activity and thus promoted higher order thinking. The study indicated that teachers of science, as opposed to other subjects, tended to use writing activities that focused on factual information rather than the underlying concepts.

While Langer and Applebee did not attempt to explain how writing was able to promote learning from a cognitive model perspective, they did highlight conditions that promoted learning. The results of their study showed that: (a) the more the content is manipulated through writing, the more likely it is to be remembered and understood; (b) the effect of writing tasks are greatest for the particular information focused upon during the writing; (c) writing tasks differ in the breadth of information drawn upon and in the depth of processing of that information that they invoke; and (d) if content is familiar and relationships are well understood, writing may have no effect at all (p. 130-131).

In summary, this study was important in assisting our understandings of how writing can promote learning because it identified a number of cognitive processes that are used during writing. It indicated how different types of writing can lead to different cognitive and conceptual outcomes. The study indicated how teachers in different subject areas place different importance on the use and value of writing and, thus, the value of writing is not uniform across the curriculum.

The third major publication in the 1980s was that by Bereiter and Scardamalia (1987), who also proposed a model to explain how writing is a learning tool. As indicated above, this model was also based on writing as problem-solving. In part, their model confirmed the ideas put forward by Langer and Applebee in that they recognized that not all writing will lead to learning. They suggested that there are two distinct types of writing, knowledge-telling and knowledge-transforming. In essence the knowledge-telling model is viewed as a recall process and does not involve the transforming of existing knowledge. The reason no knowledge transformation occurs is because "contents are retrieved on the basis of their organization in memory and the discourse schemas stored by the writer and are then translated directly into text" (Tynjala, Mason & Lonka, 2001). This model re-affirms Langer and Applebee's final point that if the content is familiar and relationships are well understood then writing may not have any effect, that is, writing does not alter the existing knowledge.

However, the knowledge-transforming model proposed by Bereiter and Scardamalia clearly pointed to an attempt to explain the cognitive processes involved with writing as a learning tool. As explained above, they adopted a problem-solving approach, but unlike Flower and Hayes they focused more attention on both content knowledge and rhetorical knowledge. They believed that each writer in undertaking the task has to initially engage with understanding the dimensions of the task by setting goals and sub-goals for completing the writing. Achieving the goals means that the writer has to deal with both content knowledge associated with the topic and discourse knowledge required to complete the writing task. Each of these knowledge bases has associated with it a "problem space", that is a content problem space and a rhetorical problem space. The interaction within and between these problem spaces is where knowledge transformation occurs. As a writer engages the content problem space and retrieves previous knowledge, then the results of this interaction will impact on decisions made within the rhetorical problem space. Within the rhetorical problem space, the decisions about how to best represent this knowledge in terms of the appropriate language for the intended audience will result in the structuring of text to reflect the writer's understandings. The results of decisions made within this problem space will inform the content problem space for further refinement of understandings where further decision making will impact on refinement within the rhetorical problem space. Thus the knowledge transforming process involves a cycling between two problem spaces with the final act being one of knowledge telling, that is, a product informing the audience of the writer's understanding.

Keys (1999) provided a description of how the use of the knowledge-transforming model can explain learning associated with using science writing tasks such as laboratory reports. For example, within the rhetorical problem space the writer not only has to focus on the language choices that they wish to use, but must also deal with "constructing canons of argument" and the need to "develop knowledge of scientific genre and the nature of science" (p.121). Within the content problem space, the writer has to "deal with relevant data", "determine meaning from the data", "construct inferences" and "develop conceptual knowledge structures" (p. 121). The results of interaction between these problem spaces is for the writer to "exercise the most important scientific thinking strategies that are associated with inquiry, the connections between problems, methods, observations, patterns, data as evidence and knowledge claims" (p.122). As writers communicate their understandings of science to an audience, they need to engage with all these demands, and thus, are cognitively active within the knowledge transforming process. Adopting the knowledge-transforming model within science classrooms involves moving past the function of writing for recording of factual information, as noted by Langer and Applebee, to focus more on conceptual understandings.

In summary, Bereiter and Scarmalia's model, particularly the knowledge-transforming model, continued the belief that writing involves some form of cycling between content and rhetorical elements. While writing was seen as a problem-solving approach to achieve stated goals, they highlighted the importance of content knowledge being strongly influenced by the manner of its representation to an audience. The function of representation engages the writer in having to deal with content and rhetorical knowledge in a manner that promotes transforming knowledge into something different from the original version the writer had experienced.

Near the end of the 1990s, Galbraith and Torrance (1999) argued for a change to the analogy for writing as problem-solving. They argued that the model proposed by Flower and Hayes placed emphasis on "effective writing as a matter of finding the right solution to rhetorical problems, and identified writing expertise with the possession of a set of sophisticated problem-solving strategies" (p. 2). The development of rhetorical goals guided the retrieval of content, and the expert was successful in both understanding the goals and the content required to complete those goals. Bereiter and Scardamlia's model was also viewed in a similar manner. The knowledge-transforming model "involves developing an explicit representation of the rhetorical problem as hierarchy of goals and sub-goals, and requires the active transformation of content in order to satisfy goals" (p. 3). The problem for Galbraith and Torrance is that the problem-solving approach portrays a controlled, rational approach to writing where skill "is a matter of deliberately working out what one wants to achieve with the text, and then working out step by step how to do it" (p. 3). They argued that a different model was needed to capture the messiness and unpredictability of the thinking processes used at different stages of writing.

Galbraith and Torrance created a new analogy for writing by choosing the phrase, *writing as text production.* Three central features for this analogy were put forward to support its use. The first was that this analogy allows for a "focus on the variations in thinking at different points within the writing process, rather than exclusively on variations in the overall quality of thinking of different writers" (p.6). In other words, they argued that a theory of how writing is produced needed to engage with different thinking processes at each stage rather than a global view. The second point was that such an analogy focuses attention on the text itself, rather than viewing the piece of writing as an "unanalysed whole, varying in length, quality or readability, but otherwise contributing relatively little to analysis" (p. 6). In other words, analysis is needed to engage with specific textual features. The third point was that the analogy "foregrounds language processes as a direct determinant of the text that is produced" (p. 6). While Galbraith and Torrance agreed that text production is partly determined by top-down processes, they claimed that it is "mediated directly by language specific processing" (p. 6). Using this analogy means that writing is viewed as shifting away from a controlled, rational process centered on explicit thinking, to one in which the "fundamental components are implicit and outside of conscious control" (p. 7).

From this analogical base Galbraith (1999) proposed a cognitive model centered on writing as a knowledge-constituting process. As with previous models proposed, Galbraith identified two knowledge bases, that is, content knowledge, which he terms the *writer's disposition*, and rhetorical knowledge, which he terms the *linguistic network*. For Galbraith (1999) when a writing task is given to a learner, then his/her content knowledge base (writer's disposition) is activated, that is, knowledge from episodic memory is recalled. In order for this knowledge to become text in the form of a written utterance, then there is a need for the activation of the writer's linguistic network (rhetorical knowledge). Having written text in front of them, which may consist of only a phrase, allows the learner to feedback through the system again to produce further writing. This feedback loop can only start when there is text in front of the leaner. Each time the learner cycles through this feedback loop, the fewer connections within their writer's disposition will be activated until a stable state has been reached, that is, no more activation of the writer's disposition occurs. This process of producing a sentence means that the knowledge originally stored is now represented differently because some meanings have been omitted in the process of deciding what to include and some were lost in the production of the sentence. This means that the knowledge contained in the text has been constituted in a new form to that which was originally activated.

As the writer continues to construct text, there is a dispositional dialectic occurring between the writer and the text, which is viewed by Galbraith as the essential element for constituting new knowledge. As such, this dialectic is constrained by two factors. The first factor is the writer's knowledge, with three elements being fixed. These are: (a) the conceptual complexity of the writer's disposition, that is the complexity between the units of content knowledge in memory; (b) the range of units activated within the writer's disposition, that is, the number of units which are actually turned on during the writing process; and (c) the linguistic knowledge of the writer, that is, the writer's rhetorical skill. The second factor is the writer's strategy for translation and is determined by three elements. These are the type of planning, the form of output, and the writer's goals.

For Galbraith, writing is a process that produces new knowledge because of an interaction between the writer's disposition and their linguistic network. It is constrained by a number of factors that determine the amount of cycling through these interactions. From this viewpoint, the optimum cognitive conditions for writing to produce new learning would be where the writer has a strong conceptual knowledge base, is able to activate a wide consideration of dispositions, has good linguistic skills, and is a good planner and goal setter. All elements would be considered crucial in effective learning from writing.

The 1990s also included a critique of previous hypotheses for writing to learn by Klein (1999). While Klein's (1999) views of hypotheses of the writing process do not completely match Galbraith's perspective, he does provide clear overlap in terms of some of the cognitive actions taken by writers. In his recent comprehensive overview of the dominant explanations of how writing enhances learning, Klein (1999) claimed that these hypotheses could be categorized into four broad groupings, each of which also theorized the role of planning. According to Klein (1999), these four groupings are: (a) writing as "spontaneous utterance"; (b) the "forward search hypotheses"; (c) the "genre-relatedhypotheses"; and (d) the "backward search hypotheses" (p. 208-209). Klein attributes spontaneous utterance to Britton (1982), who argued that student writing shapes thought in the act of expression, making tacit understandings more explicit. This is the view that the act of writing helps a writer to discover their initial thoughts on a topic. The forward search hypotheses were attributed to Bruner (1966), Emig (1977), and Donald (1991). These theorists asserted that writers transform their ideas by ongoing analyses of their texts in terms of expanding inferences, reviewing idea development, noting contradictions, and making appropriate revisions. Knowledge is constructed as a consequence of writing the text and responding to its emerging meanings.

The genre-related learning hypothesis, attributed to Bereiter and Scardamalia (1987), Halliday and Martin (1993), and Hayes (1987) among others, maintains that the use of different genre frameworks and knowledge of the micro and macro structure of texts enables students to identify the relationships between ideas, and hence clarify understanding of content. In this case by understanding the features of the genre very well, knowledge construction is achieved by application of the genre to the specific content. In other words, the genre demands of the text provide the scaffolding for new learning, as they force the writer to organize and understand relationships among textual ideas.

The backward search hypothesis, which Klein attributed to Flower and Hayes (1980; 1984), Vygotsky (1962) and Bruner (1966) among others, argues that students learn through writing by setting and addressing rhetorical and content goals. Knowledge is constructed by the setting of a goal, and then completing a series of sub-goals as a means to achieving the overall goal of the task. In assessing the relative coherence of each hypothesis and the research evidence, Klein (1999) concluded that each theory offered only "probabilistic rather than certain" (p. 254) evidence of learning through writing. Klein noted that there was some research evidence for the value of genre-based writing; however, he emphasized that in order for this writing to promote student learning, students must know and seek to adopt the goal of a given genre, have strategies to achieve that goal, and in this process transform their content knowledge into new learning. For the author, there is clear overlap between the forward and backward search hypotheses put forward by Klein and the work of Galbraith. The initial cycling through the interactions of the writer's disposition and the linguistic network represents elements of the forward search hypothesis, while the limiting factor of the writer's goals is represented in part by the backward search hypothesis.

SUMMARY

In response to the call for a theoretical model to explain the cognitive processes associated with writing as a learning tool there are a number of features that are in common agreement between the models put forward since Emig's original paper, in what has been a cumulative theory-building pattern. Importantly, recognition is given to the concept of interaction between an individual's prior content knowledge (or disposition) and their discourse knowledge (or linguistic network). While the mechanism of the interaction is not clear, there is a dialectic which occurs between these forms of knowledge that is controlled both by the writer's level of expertise and level of understanding within the respective knowledge bases. Recognition is given to the range of constraints which impact on the construction of text, and hence the meaning which is constituted. While writing is more than just problem-solving, the skill and knowledge base of the writer does impact on the quality of the constructed text.

Another important consideration that was highlighted by Keys (1999) in her description of using Bereiter and Scardamalia's model, is the constraint of the nature of the discipline when using writing as a learning tool. Science is a discipline that is based on particular patterns of argument incorporating the need for claims, evidence and warrants. As such, any writing used within the field of science needs to ensure that there is adherence to what constitutes the essence of the discipline. This means that the writer not only has to engage with the nature of science constraints within their knowledge base, but also within their linguistic network.

CHAPTER 4

THE ROLE OF LANGUAGE IN SCIENCE LEARNING AND LITERACY

Vaughan Prain

There has been increasing recognition in recent years of the central role of language in both learning science and in developing science literacy, where science literacy is seen as the desirable general outcome of learning science. At the same time, the definition of science literacy has been broadened over the last decade from a traditional focus on technical conceptions and terminology to include skills in communicating science concepts and applications. The American Association for the Advancement of Science (1993) and the National Research Council (1996) emphasized the need to develop students' cognitive abilities, reasoning, habits of mind, unifying concepts, and communication skills. The National Science Education Standards (NRC, 1996) asserted that science literacy entailed the development of educated citizens who can "engage intelligently in public discourse and debate" (p. 13), can "construct explanations of natural phenomena, test these explanations in many different ways and communicate their ideas to others" (p. 20). In supporting this general orientation, Hand, Prain, Lawrence, and Yore (1999) asserted that an explicit focus on using language for communicative purposes was crucial to achieving this broadened account of science literacy.

Over the last fifteen years there has also been strong interest in the role of language in learning science. This interest parallels Christie's (1981) account of how language and learning are connected generally. Christie (1981) claimed that learners can learn through a language, learn about a language, and also learn how to use a language. In the first relationship, or learning through a language, language is understood as a resource for understanding new content, where use of everyday words and concepts enables learners to explore and clarify the meaning of technical terms or new concepts. In this sense, the learner's current language competencies provide a means to develop new understandings, new vocabulary, and new language practices. In the second relationship, or learning about a language, learners study a language as a system, as in learning the history, linguistic structures, grammar, logic, components, and values of a language. This learning may also include understanding possible limitations to these values and their usage. From this perspective a language is always part of a culture, and reflects aspects of that culture. Christie also noted that every language has variations in expression that are still recognizable as examples of that particular language. In other words expression in a language is dependent on context, purpose, readership, and the norms of usage in different settings. The third relationship, or learning how to use a language, is concerned with learning how to speak, read, write, and view the target language competently in varied contexts for different purposes and audiences.

Christie proposed that each of these three relationships was interdependent and that learners needed to be proficient in each area to be considered competent language users, or literate. Researchers in science education, over the last fifteen years, have advocated strongly the need for each of these possible relationships to inform student learning of science, where science is understood, among other things, as a particular language, or more recently, languages, with a history, values, discourses and rationale.

Most scholars agree that student talk is valuable for promoting each of these three areas of learning, where students' everyday language, guided by their teacher, is seen as an important resource for learning new science concepts and practices (Halliday & Martin, 1993; Lemke, 1990; Ogborn, Kress, Martins & McGillicuddy, 1996; Sutton, 1992; Wellington & Osborne, 2001). Many researchers have noted routinely that the scientific and everyday meanings of some words differ, and hence cause problems for using talk for learning. However, students' everyday language and ontological beliefs (their understandings about the natural world and what exists) and their epistemological beliefs (their ideas about the basis of this knowledge) need to be engaged, and are a necessary starting point in learning new concepts, terminology, and the methods of science. Many scholars also posit that students should learn how to talk the language of science, in terms of appropriate vocabulary and methods of scientific reasoning and inquiry.

TWO VIEWS OF WRITING TO LEARN

On the question of the relationship between students' written language and learning science, researchers have divided broadly into two perspectives about which practices should be foregrounded and on what basis. One approach, focusing on learning the language of science as a system, asserts that students primarily must learn to understand and reproduce the traditional written discourses of the science community if they are to become scientifically literate (Halliday & Martin, 1993; Unsworth, 1997, 2001, Veel, 1996, Kelly & Chen, 1999). This approach assumes that science language is a particular set of language practices, and that students need to master these ways of doing and recording science. This "learning how to write science" approach has been justified mainly on epistemic or knowledge grounds in that it is viewed as consistent with how knowledge is generally claimed, organized and represented in this subject area.

A contrasting approach, drawing generally on pedagogy-based justifications, and consistent with Christie's (1981) view of language as a resource for learning, asserts that for various reasons students should use a more diversified range of writing types, both formal and informal, to acquire science literacy as well as knowledge of, and particular attitudes towards, scientific inquiry (Boscolo & Mason, 2001; Hanrahan, 1999; Hildebrand, 1998, 1999; Hodson, 1998; Prain & Hand, 1996a; Stadler, Benke, & Duit, 2001; Sutton, 1992; 1996). The diversified writing approach assumes that written language can be a valuable resource for various aspects of learning science and science literacy, and proposes a broader role for writing than mainly as a record of past learning or as an imitation of the 'official' writing formats of practicing scientists. Advocates of the diversified approach, such as Lemke (2002) also claimed that students needed alternative discursive resources and ideological perspectives outside the values of mainstream science if they were to critique the effects and ideological implications of science practices. Both broad orientations have strengths and weaknesses in terms of their current claims, proposed classroom practices, and relationship to new technologies; and there are various compelling arguments to justify student engagement with both kinds of writing. This chapter reviews the claims and counterclaims of both perspectives, then goes on to consider recent research on science as a multi-modal subject, entailing the use of graphs, diagrams, and non-linguistic communication. The future role of writing in science in schools is discussed in relation to this research.

Learning How to Write Science

Researchers within the writing science approach generally view writing for learning as the process whereby students learn about the functions of linguistic features in traditional forms of science writing, such as the experiment or research report. Features are identified at both the micro level, including vocabulary and grammatical features, and at the macro level, including larger textual elements such as introduction, discussion, and conclusions. While other genres, or types of writing, such as biographies of famous scientists and historical accounts of science are acknowledged as legitimate examples of writing about science, the experimental report is emphasized as the key genre of scientific method. The importance of the experimental genre has drawn heavily on systemic functional linguistics, especially the work of Halliday (1978) and Halliday and Martin (1993).

Halliday and Martin (1993) asserted that the epistemic distinctiveness of science as a worldview, as a body of knowledge, and as a form of inquiry with various technical specifications, is indivisible from the development, over several centuries, of a range of purpose-built features of language use. Through analysis of various historical and contemporary instances of scientific argumentation, they argued persuasively that specific grammatical resources of English have been used to construct and represent the specialized knowledge of science, as disseminated in science communities. This writing is characterized by high lexical density, interlocking mutually dependent definitions, and subject-specific terminology to develop and enable scientific methods of inquiry and reportage of scientific results.

For example, in scientific writing verbs are often nominalized into nouns, so that specific actions and processes can be characterized as universal entities and then deployed in building elaborate causal connections or scientific explanations without the need to re-elaborate the basis of the case or evidence each time. Examples of familiar nominalized verbs include photosynthesis, metamorphosis, and crystallization. Halliday and Martin (1993) claimed that the resultant brevity of expression and density of meaning were essential attributes of writing scientifically, and therefore students needed to master and reproduce these linguistic features to be considered scientifically literate. Similarly, they argued that various genres have been developed to provide appropriate macro-structures to represent scientific reasoning, argument and discourse, and these linguistic aspects were viewed as representing the essence of science as a discipline and field of study. According to Unsworth (2001, p. 125) these genres, for the purposes of learning science in schools, included procedural recounts, explanations, descriptive reports, taxonomic reports, expositions and discussions. While Unsworth (2001) has recognized that lengthy texts often blended different generic structures and purposes, he continued to assert that student writers needed to learn the form/function relationships of distinctive science genres if they were to become scientifically literate.

Various implications follow from this focus. Science literacy is conceptualized as technical knowledge about, and ability to reproduce, a narrow range of subject-specific micro and macro linguistic features of the language of the science community. By implication, the use in the classroom of 'non-scientific' genres such as narrative or poetry, or simplified explanations of scientific topics in popular magazines, were viewed as questionable for learning science or representing scientific knowledge. Martin (1993) claimed that proposed diversified science writing tasks in Australian State science curricula during the 1980s were misguided from an epistemic perspective, patronizing to learners, and a disservice to effective science learning. Halliday and Martin (1993) also argued that scientific writing, because of its compression of meaning, differed markedly from equivalent verbal language attempting to explain scientific concepts, and therefore students needed explicit instruction in the grammatical features of scientific genres if they were to understand scientific method and science writing as a practice. In this view the specific meanings of scientific terminology and grammatical structures of science writing are not equivalent to everyday language, and while everyday language provides starting points for learning, science literacy must entail student knowledge of the exact scientific meanings of words and concepts and their appropriate usage in scientific discourse.

From this viewpoint students need to learn the assumptions, procedures, hidden rules, and purposes of scientific writing as the basis for understanding what counts as scientific method, explanation, and justification, as well as the underlying history and rationale of this writing as a practice. Knowing the generic rules, it is argued, enables pupils to "process information deeply" as they "construct relationships among ideas" (Klein, 1999. p. 230). Researchers in this orientation have focused on the discipline-specific structural and functional features of types of science writing (Halliday & Martin, 1993; Unsworth, 2001), their subject-specific vocabulary, and the student knowledge required to understand and reproduce these genres (Unsworth, 1999, 2000, 2001; Veel, 1996; Martin & Veel, 1998). According to Martin (1999), Veel (1996) and others, students will effectively learn the rules and meanings of this particular language practice through the following teaching strategies: detailed analysis of linguistic features of textual examples; joint construction of genres with their teacher; and through an explicit extensive teacher focus on key textual function/form relationships and their rationale. This approach assumes that the most effective way for students to learn science through writing is to imitate the writing practices of professional scientists, or at least a simplified version of these practices in school science genres. Unsworth (2001) argued that students can learn to write scientifically through analyzing the schematic structures and grammatical patterns of sample texts, and then reproducing these functions in their own writing.

Diversified Writing Approach to Learning in Science

In contrast to the school science generalist approach, other researchers, drawing on a diverse range of educational theorists, including Bereiter and Scardamalia (1987), Galbraith and Torrance (1999), Klein (1999), Gianello (1988), and others, have emphasized the value of expanding the purposes, writing types, and readerships for writing in science beyond induction into traditional school genres. Writing here is viewed as a resource to enable learners to understand science concepts, scientific method, and to practice some methods used by scientists in representing science. This approach also assumes that students should write to respond critically to scientific practices and their application to social and economic issues beyond the classroom. While the classroom community of science learners is clearly not the same as a science research team, in that students are not expected to generate new science knowledge for others, some practices, such as clarifying ideas through notes and diagrams, and seeking to persuade others in the team about a particular viewpoint, and disseminating ideas to diverse readerships, may overlap, and also support classroom learning.

Educators such as Boscolo and Mason (2001), Hanrahan (1999), Hildebrand (1998; 1999), Prain and Hand (1996a), Rowell (1997), Stadler, Benke and Duit (2001), Sutton (1992, 1996) and others, have asserted that students, in striving to clarify networks of concepts in science topics, should be encouraged to write in diverse forms for different purposes. Gunstone (1995, p. 15) argued that there are learning gains when students write "translations" of their emerging understandings in a variety of genres using their own words, or everyday language. Such a view draws on social constructivist perspectives of learning as the process whereby learners use and engage their own linguistic resources and conceptual categories and frameworks to demonstrate understandings in relation to target concepts and technical vocabulary (Hand & Prain 1995; Fensham, Gunstone, & White 1994). This viewpoint does not imply that learners have innate competencies that enable them to learn science vocabulary and concepts without teacher guidance, but rather that writing can be a useful resource, parallel to discussion, to clarify science concepts and practices, to connect new concepts and meanings to past understandings, and to develop critical perspectives. Student writing is here viewed primarily as a resource for thinking and learning as students clarify and consolidate conceptual knowledge. Within this perspective such writing can also promote students' science literacy by developing their interest in and capacity to apply scientific thinking to social issues for the purposes of informed action, where students learn to cross borders between specialist and more popular genres and readerships. Such writing can also develop students' strategies for self-directed learning of science, and provide positive affective engagement with science for all students, not just those who might specialise later in this subject area. From this perspective writing in science is not only a means to display knowledge of scientific understandings, procedures, processes and rationale, but also functions as an epistemological tool, as a resource for understanding and refining knowledge.

This approach assumes that students' everyday language, their broad knowledge of a diverse range of genres, and their common-sense categories about causality in relation to natural phenomena are crucial resources for developing their understandings of the technical language and counter-intuitive concepts of science. Both theoretical and practical arguments have been advanced to support this approach, including claims for a closer reciprocity between everyday and scientific language, metaphors and theories. In supporting this latter argument, Sutton (1992) asserted that scientists, historically and in contemporary research, have used metaphors from non-scientific contexts to represent or encompass their emerging theories and understandings, and that a change of metaphor always signals an altered theory.

Halliday and Martin (1993, p. 20) supported this perspective on the crucial role of everyday metaphor in shaping and changing scientific theories and explanations. They claimed of scientists, "the language they learnt at their mothers' knees is much more in harmony with their deepest theoretical perceptions." Recognition of the key role of everyday metaphors in representing scientific theories, and translating complex explanations into accessible language and concepts, would seem to justify a pedagogical focus on the role of metaphor in learning science, and the value of students using writing to explore, justify and evaluate different metaphoric explanations in science theories. Such an approach is also consistent with Sutton's (1992) concern to reinstate the human, historical and personal aspects of scientific explanations in learning science.

Rowell (1997), Keys, Hand, Prain and Collins, (1999) and Tucknott and Yore, (1999) claimed that when students are beginning to learn new concepts, there are pedagogical advantages for them to write in diverse ways that lead them to connect technical vocabulary to their everyday language and past experiences. Prain and Hand (1996a) claimed that diverse writing genres such as booklets, travel brochure, and letters to editor, not only represent ways for students to consolidate their knowledge of science concepts, but also corresponded to the variety of actual writing tasks literate adults, engineers, and some scientists use in writing about science. Michaels and Sohmer (2000) also claimed that teachers needed to provide opportunities for students to build connections between their personal meaning-making processes incorporating their everyday perceptions of the phenomenal world and the official inscriptions and representations of school science knowledge.

Research claims for the benefits of this approach have focused mainly on the positive effects on students' attitudes towards and engagement with the subject (Hand & Prain, 1995; Hanrahan, 1999; Hildebrand, 1998, 1999; Prain & Hand, 1996b; 1999). Students have consistently reported across a range of studies that this focus on making sense of scientific ideas and justifying their views were valuable to their learning, and also promoted a positive attitude towards the subject when contrasted with undertaking traditional writing tasks. These studies also reported success with students who in the past failed to engage with science, or performed poorly. Hildebrand (1999) reported that diversified writing tasks, including more imaginative writing, assisted students' learning processes, had strong motivational effects, and improved learning outcomes.

Criticisms of Each Approach

There has been ongoing debate by advocates of each approach, aimed at identifying strengths and weaknesses in each position. Advocates of the school science genrist approach have argued that learners disadvantaged by class, race or language difficulties needed additional explicit instruction in the structural features of science writing if they were to succeed in cracking this subject's 'codes' and rules. Educators such as Champagne and Kouba (1999), Martin and Veel (1998), Scheppegrell (1998) and others claimed there were learning gains for this approach over progressive pedagogies that focused on more student-initiated learning because tacit knowledge was made explicit. Martin and Veel (1998) asserted that progressive pedagogies where students were expected to pose problems, ask questions and negotiate the focus of topics favoured middle-class, knowledgeable, confident, motivated learners, but failed to provide a successful science learning environment for disadvantaged students. Other researchers, such as Hand and Keys (1999) have argued that students benefited from engaging with modified versions of traditional genres, where the new framework encouraged students to be more explicit about their reasoning processes in responding to prompts for each component part of the experimental report genre.

Researchers within the genrist perspective have questioned the value or feasibility of attempting to 'translate' understandings from technical to everyday language, or into different genres. Martin (1993, pp 234-5) argued that "the process of translating common sense into specialized knowledge is ... one with an indefinite number of steps intervening between a technical term and the every-day 'core' vocabulary through which it might eventually be rendered to an absolute novice". This could imply that the task is unreasonably demanding and/or a distraction from 'real' science writing, namely student mastery of the authentic genres of school science.

Translation theorists such as Whelan (2002) have also acknowledged that there is never absolute equivalence of meaning between different forms or expressions of a language, and therefore there is always a question of degree of match between expressions, rephrasing, of standards of equivalence in any translation or rewording. However, the purpose of the translation activity or elaborated explanation is not for students to find exact synonyms for technical terms, but rather for the process to enable students to contextualize understandings in richer associative networks or elaborated contexts, as opposed to rote learning and regurgitation of technical terminology. Clearly Martin's argument is persuasive if all science concepts and reasoning processes were expected to be unpacked in this way by students in writing in this subject.

However, research on students' attempts to reword and explain some concepts within topics in less technical language (Prain, Hand & Hohenshell, in press) especially complex interlocking concepts that do not lend themselves to routine classroom practical and observational work, has indicated some positive learning gains from this process. Halliday (1993, p. 70) also claimed that it is not possible to "represent scientific knowledge entirely in common sense wordings" because "the conceptual structures and reasoning processes of physics and biology are highly complex and often far removed, by many levels of abstraction, from everyday experience". However, there is still a need to identify which writing tasks and topics, including computer-based tasks, supported by instructional programs facilitate effective learning of target concepts when student attempt to link the concepts to broader or less self-referential frameworks.

By contrast, advocates of diversified genres, such as Hildebrand (1999), Sutton (1992) have cited a range of studies of famous scientists reporting on the central importance of non-scientific metaphors in new scientific conceptions, and the use of imagination, speculation, feelings and intuition in science method and creativity. By implication, some student writing in science should provide opportunities for addressing these aspects of scientific inquiry. In elaborating this case, Hildebrand (1999) argued that traditional science writing advocated in schools, with its focus on rational, objective, factual and abstracted knowledge, failed to capture the documented richness of scientific investigation and writing. She further asserted that there was now a substantial body of literature discrediting the traditional inductive scientific paper as an accurate representation of the work of scientists, and substantial evidence of the ways in which science writers used rhetoric, metaphors and literary devices to make their findings convincing to readers. She also asserted that a restricted focus on school science genres failed to acknowledge the many forms of informal and formal writing undertaken by scientists in developing their ideas, including notes, jottings, posters, letters, and email messages. Research by Hand, Yore, and Prain (1999) also confirmed a richer pattern of actual writing by scientists than proposed by a narrow genre approach. Lemke (2000, p.300) claimed that science was falsified as a study when concepts were taught outside their "social, economic, historical, and technological contexts." In other words, the decontextualized representation of concepts in school science genres failed to address factors influencing their formation and application to social issues.

Educators from feminist perspectives, such as Haraway (1989, 1991) Hildebrand (1999), Shepherd (1993), Keller and Longino (1996) and others have argued for the need for both critique of the narrowly masculinist, rationalist account of science implied in current writing practices, and the opportunity for learners to write about the experiential, aesthetic aspects of scientific observation, concepts, and understandings. Supporting this argument, Brickhouse (2001), Lemke (2001) and others claimed that traditional genres failed to engage with the affective, embodied aspect of student learning of science, and researchers such as Aikenhead et al (1999), Lee and Fradd (1998), and others have questioned the appropriateness of traditional generic writing for learners with diverse social and cultural backgrounds.

Value of Both Approaches in Writing for Learning

Both approaches have developed valuable insights into the relationship between different kinds of writing and learning in science. The genrist approach has identified complex language challenges facing students in mastering the different genres of science, as they become enculturated into scientific discourse. This strand of research has also identified some effective pedagogical strategies that can facilitate this learning. However, if students are to learn to write like scientists, then it is important that school genres match the actual writing purposes and practices evident in the writing of professional scientists. This suggests that school science genre writing should be responsive to changes evident in new technology-based multi-modal representations of science, and should also take account of the emerging diversity of forms of scientific inquiry now evident in multi-disciplinary science research teams. In this way, the writing of school science genres can be an appropriate induction into how science is practiced and represented in both the science and the broader community.

The diversified writing approach has identified some effective communicative contexts and frameworks for enhancing students' engagement in, and understanding of, science concepts and practices, and also opens up opportunities for student to critique science goals and effects on society. However, while there is emerging evidence of beneficial learning and attitudinal effects in diversifying students' science writing to consolidate understandings of concepts and science practices (Hand, Prain, & Yore, 2001), various issues still remain to be explored in order to maximize the positive outcomes of this approach. These issues include identification of the best match between topic, genre, target concepts, and different readerships. There is also a need for systematic identification of those analogies, models, metaphors, and communicative contexts that are most conducive to students learning through more diversified kinds of writing. There is also a further need to identify which teacher practices and orientations, classroom processes, specific writing tasks, and evaluative processes, best support this approach. Such a research agenda should also identify the effects of these practices on different groups of learners, including those not usually considered successful in learning science.

WRITING AND MULTI-MODAL LEARNING OF SCIENCE

The recent research interest in writing for learning in science also parallels a growing recognition of the key importance of understanding and conceptually integrating different representational forms in learning science and learning how to think and act scientifically (Ainsworth, 1999; Dolin, 2001; Lemke, 1998; 2001, 2002b; Roth, 1995; Russell & McGuigan, 2001; Stenning, 1998). These researchers have asserted that to learn science effectively students must understand different representations of science concepts and processes, and be able to translate these into one another, as well as understand their co-ordinated use in representing scientific knowledge. While various classifications of these modes have been proposed, there is broad general agreement that these forms, for the purposes of learning science in secondary school, include such categories as descriptive (verbal, graphic, tabular), experimental, mathematical, figurative (pictorial, analogous and metaphoric) and kinaesthetic or embodied understandings or representations of the same concept or process. There is consensus in the literature that students need to develop an understanding of diverse modes, rather than be dependent on particular modes for specific topics, if they are to develop a strong understanding of how to use and represent science concepts. There is also agreement that these modes of representation have different strengths and weaknesses in terms of precision, clarity and associative meaning. As noted by Lemke (2002a), 'drawing and gesture are much more powerful at expressing topological, and therefore quantitative meanings, while verbal language is much better at reasoning about relations among categories'. Students need to understand these aspects of the representations, as well as their integrated use to represent scientific concepts, especially in upper secondary and tertiary levels of studying science.

At more junior levels of science learning, students need to be introduced to multiple and multi-modal representations of science concepts and be able to understand, translate and integrate these theoretical representations as part of an introduction to learning the nature of scientific knowledge and inquiry, and its representation. Current science teaching practices incorporate the use of both authorized or justified representations as well as student-generated multiple and multi-modal representations for some topics (such as the use of 3D models, diagrams, verbal accounts, role-play, and CD-rom illustrations for teaching topics like the solar system). At the same time there is a growing recognition of the need for students to learn how to interpret, integrate, and reproduce these multi-modal representations within and across topics if they are to develop robust science literacy.

Dolin (2001) and Roth (1995) observed that various representational forms have been under-utilized in learning science and could be effectively incorporated into classroom practices, assuming that learning is enhanced where students integrate different representations of the same concept or process. In using this general approach with primary school students, Russell and McGuigan (2001) argued that learners need opportunities to generate a variety of representations of a concept, and to recode these representations in different modes, as they refine and make more explicit their understandings. In their classroom program both student and teachers generated various representations of target concepts, and knowledge construction was viewed as the process of making and transforming these different representational modes as they scaffolded their understandings in relation to their perceptions of the real world. Dolin (2001) noted that senior secondary physics students achieved enhanced understanding of concepts in physics when they attempted to translate different representational modes into one another in that subject. He further argued that this border crossing between modes was essential for developing strong conceptual understandings. By implication, this view of the multiple representational languages of science (visual, verbal and mathematical) suggests a modified role for written language in the learning process. From this viewpoint written explanation is seen as only one mode of conceptualization, and increased emphasis might be given to integrating modes or focusing on non-verbal modes. Supporting this view, researchers such as Gobert and Clement (1999, p. 49) have claimed that students can 'draw to learn' effectively, where the visual media affords 'specific advantages over the textual media' (p. 50).

IMPLICATIONS FOR FUTURE PRACTICE

This increased recognition of the multi-modal nature of science clearly raises a range of challenges for understanding the role of future student writing for learning. There is the need to ensure that student writing tasks parallel the emerging diversity of practices of professional scientists as they carry out scientific activity. There is a need for a comprehensive account of \these diverse science practices, so that science education in schools, including writing, can function as a genuine "semiotic apprenticeship" (Hodson, 1998, p. 123), modeled on the diversity of actual professional science activity. The increasing range of different genres for engaging in, and representing science on the Web, as noted by Lemke (2002a), confirms how much genres are indeed 'open systems' (Martin, 1999, p. 141), and indicates the need for future flexibility in teaching the forms and functions of science genres in schools. As Christie noted (1981, p.3), the 'varieties' of a language need to be recognized as part of what counts as the content and skills required for effective learning. This implies that students need to understand and be able to generate the diverse range of ways in which science is practiced, reported, and cited, including popularized accounts of science for varied readerships. A further challenge will be to identify the new role of writing within more complex, multi-modal representations of science inquiry, where an understanding of the strengths and limitations of this mode is needed to guide science teachers' choice of writing tasks. There is also the further challenge of developing students' capacities to evaluate critically in writing the purposes, claims, and credibility of science methods and findings.

CHAPTER 5

A RESEARCH PROGRAM ON WRITING FOR LEARNING IN SCIENCE, 1992-2002

Brian Hand & Vaughan Prain

This chapter reports on the development, implementation strategies, and broad findings of a long-term research program (1992-2002). The program aimed to investigate the effects on learning outcomes and attitudes for students and teachers when students wrote in diversified forms to learn science. In the context of this program we defined science learning very widely to include scientific methods, concepts, and various forms of communication. However, the main focus has been on the understanding of science concepts and their application in different problem-solving contexts.

Beyond our central question on the role of writing for developing conceptual understanding, we also sought to address a range of connected questions relating to effective classroom practice. In attempting to clarify aspects of effective practice, we examined the following issues: In what ways exactly does diversified student writing support learning of concepts? What conditions are necessary to maximize the potential of student writing to support this learning? The conditions that the researchers investigated included task design, features of classroom learning environments, teacher practices in setting up, implementing, and assessing student writing, psychological conditions relating to student engagement with writing tasks, and student knowledge of appropriate strategies for generating, revising and learning from their text production and revision. The major focus of the research from the outset was to develop a pedagogically sustainable approach to enhancing the potential of writing to promote student learning in science.

The research program has fallen broadly into two phases, reflecting a shift in focus from initial exploration of implementation issues to broader quantitative studies on the effects on students' learning through sequenced, richly contextualized writing tasks. The first phase (1992-96) entailed small-scale qualitative studies of classroom environments, focusing on teacher and student practices in using diverse writing genres to represent science concept and methods. This phase was concerned with identifying how a small group of teachers engaged with implementation issues in diversifying student writing tasks, and the effects on students' understandings. The second phase (1997-2002), building on the positive outcomes of the first phase, has sought to identify optimum conditions for student writing to serve learning, as well as to quantify the learning gains of this approach over more traditional writing tasks. This has led to an increased focus on multiple and cumulative writing tasks, planning-for-writing strategies, and effects of writing for different readerships and communicative contexts on conceptual learning.

The research program began in the context of the increasing acceptance of language as a crucial medium of learning across the curriculum. This emphasis on language as a resource for understanding entailed both transference of methods for learning to write in English classes to other subjects (Fulwiler & Young, 1982; Martin, 1984) and identification of the particular demands of writing in the discourses of different subjects (Fulwiler, 1987; Walvoord, 1982). At the same time, educators such as Shymansky and Kyle (1992), Holliday, Yore, and Alvermann (1994) and others advocated the value of cross-disciplinary studies in education research so that insights and strategies from different subjects could be consolidated to improve learning through language generally. As two researchers with different backgrounds in science education and literacy education, we were interested in a cross-disciplinary approach to help characterize the role of written language in developing, displaying and assessing students' understanding of science concepts and methods. While some research had been conducted on the role of talk for learning, the role of written language in consolidating student learning of science was relatively under-researched at that time.

In the early 1990s, a survey of seventy science teachers in the Australian region of Bendigo indicated that writing assignments for students generally a narrow range of purposes. These purposes were dominated by note-taking and short answers to questions (Hand & Craker, 1991). There was little variation from this pattern apart from some teachers also setting students to undertake relatively independent research topics or project work. On the basis of a subsequent inquiry by a smaller group of 10 secondary science teachers in Bendigo in 1992, an inservice program with the two researchers was proposed to focus explicitly on the role of language in learning science. In initiating this program the researchers were concerned to (a) identify teachers' current understandings of the role of language in learning science (b) devise, trial and evaluate a range of writing tasks to promote student learning, and (c) to develop participant teachers' knowledge and pedagogical understandings in shifting to a broader use of writing types for learning in science.

THEORETICAL FRAMEWORK

In the early 1990s there was widespread recognition in Australian national and state curriculum documents of the need to change the role of language in teaching and learning in subject areas (DEET, 1989; Parker 1992). Drawing on the research of Bruner (1964), Barnes (1976, 1986) Emig (1977), Scardamalia and Bereiter (1986) and others, this perspective on language for learning advocated a shift from text-and-teacher dominated classrooms to more open-ended, speculative language uses by students. Within constructivist orientations in science education (Fensham, Gunstone & White, 1994; Hand & Prain, 1995) teachers were encouraged to change from a transmission practice to a more interactive teaching style that acknowledged and drew out students' diverse conceptual frameworks. There was general agreement that constructivist teaching approaches should entail strategies that recognized and responded to this diversity. This orientation implied that students needed diverse opportunities to clarify their understandings of key science concepts and be able to demonstrate their application in multiple contexts. From this perspective, in traditional classes, students' understanding of the concepts and technical vocabulary of science was under-contextualized in terms of rich networks of explanatory meanings. Our research project sought to address this limitation through using writing as a tool for fostering these semantic links.

There was also a general research consensus in the early 1990s that "different kinds of writing tasks eventuate in different kinds of learning" (Rivard, 1994, p. 978; Schumacher & Nash, 1991). One response to this within constructivist perspectives was to recommend that students write in a range of types for a variety of purposes to initiate new thinking and consolidate learning as they construct new understandings (Shymansky & Kyle, 1992; Sutton, 1992). As Rivard (1994) in a review of research on writing for learning in science noted, there were additional interconnected factors that influenced the success of this kind of learning. These were: (a) the learner's understanding of strategies to guide the writing; (b) the contextual aspects including a classroom learning environment focused on deeper conceptual understandings rather than factual knowledge; and (c) a match between genre or type of writing, conceptual structure of the topic, and broader curricular goals. Any effective change to students' writing practices in science for learning therefore entailed a complex set of interlocking conditions, where teachers and students needed to understand clearly the goals, methods and success indicators for using writing for learning. Teachers needed to be clear and confident about how to set up, support and assess these new practices, and students needed to understand how and why they should monitor their own learning strategies.

The researchers also recognized that current assessment practices at that time tended to emphasize factual recall over deeper conceptual knowledge. They reasoned that assessment methods might need to be altered to identify the expected different learning outcomes from an increased focus on the role of writing to promote conceptual development and application to new contexts. To engage with the potential for writing to support this kind of learning, students needed to tackle tasks that required them to elaborate, and reprocess concepts and central ideas. They needed to hypothesize, interpret, synthesize and persuade, and hence develop higher order thinking skills (Resnick & Klopner 1989; Schumacher & Nash, 1991; Sutton, 1992). They needed to demonstrate that they could transfer declarative knowledge about science concepts to new contexts as part of higher order thinking. Minstrell (1989) and Roth (1990) noted that the process of conceptual change for students was strengthened when they reflected upon their alternative conceptions, responded to current authorized explanations, and considered real-world applications within a supportive classroom context that tolerated multiple interpretations.

In this project student writing was conceptualized as a complex problem-solving and knowledge-building process that was influenced by many interlocking factors including the following: (a) the students' conceptual knowledge of and attitude towards the topic; (b) their understanding of generic requirements of tasks; (c) their perception of the demands of particular readership needs; and (d) an understanding of the methods used to produce and revise texts. Consistent with contemporary research on process approaches to language learning (Flood, Jensen, & Squire 1992), the researchers assumed that students needed to experience a strongly supportive learning environment if the increased effort of extended engagement with written texts was to lead to effective student learning as well as positive attitudes towards these new practices. Such a supportive environment was conceptualized as including opportunities to vary how student writing was produced, including individual, partnered and group drafting, as well as exchanges of drafts, and an environment where learning was richly contextualized in terms of science activities, diverse discussion opportunities, and an orderly purposeful sequence of informative feedback on emerging student work. Student writing was to be conceptualized as an ongoing problem-solving space, a site for work-in-progress on understanding key concepts in science, rather than one-off displays of understanding or memorization.

THE FIRST RESEARCH PHASE: IMPLEMENTING WRITING TO LEARN SCIENCE IN THE CLASSROOM

In designing the first research phase, the researchers drew heavily on constructivist philosophies (von Glasersfeld, 1989). These emphasized participatory decision-making rather than transmission delivery of packaged programs by an "authority" to "novices". Also, we recognized the crucial role of varied individual perspectives in effective sustained learning. There was no assumption in the program that the teachers would be compelled to reorganize their teaching practices or compulsorily reconceptualize their sense of writing for learning in science. A major guiding assumption of the program was that the implementation and institutionalization of reform initiatives could best be achieved through a cooperative partnership between researchers and teachers working at a local level over extended time. The program was designed around exploring teachers' own perceptions of language use within the classroom, providing opportunities for discussion of conceptions of assessing and learning from students' written work as outlined from an English educator's viewpoint, the trialing of new approaches, and group-based reflection on the success of implementation of these new approaches. The researchers were keen to establish an environment in which teachers were able to reflect actively on, and construct new conceptions of, language use.

Context and Methods of the First Research Phase

The initial studies took place in five junior secondary colleges (years 7-10) in Bendigo, Australia, from 1993-1996 with a group of ten volunteer secondary science teachers who formed part of an ongoing in-service program. Bendigo is a regional centre of 65,000 people with a diverse socio-economic profile, where students in the program were predominantly from Anglo backgrounds, with a small proportion of Asian and Pacific Island students. The teachers in the study varied in teaching experience from three to twenty years, with one teacher having a master's degree in science education. The program was intended to provide an environment in which teachers were able to reflect actively on their current understandings of the role of language in learning science, and to devise, implement and evaluate a variety of writing tasks that could be incorporated into their normal science programs. Each session, apart from the initial one, began with teachers reporting to the group as a whole on their progress with the new writing tasks. The program consisted initially of a series of six half-day meetings held every three weeks, with teachers being released from classroom duties during the second semester of 1992.

A qualitative research method grounded in constructivist epistemology (Lincoln & Guba, 1985) was adopted for this project because such a method reflected the researchers' concerns to focus on participants' perceptions of successful pedagogy in learning through writing. The researchers considered that in order to investigate teacher and student perceptions it was necessary to undertake small-scale studies of classroom practices. A case study approach (Merriam, 1988) was used where participant teachers were encouraged to explore, reflect on and construct their own conceptions about the use of writing approaches within the classroom.

To enable an accurate record of this process four data sources were used throughout the project, journals, semi-structured teacher interviews, student interviews and records of student work. Teachers were asked to keep a journal throughout the in-service program to record their conceptions of material presented and discussed during each session, their impressions of the implementation of the new approaches within the classroom, and the reaction of students to the changes from previously adopted teaching strategies. Semi-structured interviews were held on completion of the in-service program focused on such issues as teachers' conceptions of the program prior to and on completion of the program, the student inclusiveness of the new approaches trialed, teachers' perceptions of science given the implementation of different pedagogical approaches, and directions for the future. Each teacher was asked to collect samples of students' work to bring to the discussion sessions held at the beginning of each in-service session. Teachers were asked to comment on these products, with discussion points focusing on the type of students who had completed the work, the involvement of these students in science generally, the value gained for the students and the teacher. Records of these discussions were kept.

Each participant teacher trialed new strategies in at least two of his/her classes, and from the approximate number of 500 students a sample of interview responses from 50 students was obtained. These students were selected at random but based on guidelines relating to ability range and gender balance. Ability range was defined as a three-point scale of weak, average or above average based on previous science assessment. Questions in these interviews focused on students' conceptions of the value and appeal of the new writing tasks both for themselves and in terms of learning. A further issue focused on their perception of changes to classroom procedures resulting from these new forms of assessment.

Outcomes of the First Research Phase: Teacher Implementation and Student Response

The teachers selected the classes and topics they wished to use for the project, resulting in a diverse coverage of topics across different year levels including electricity, genetics, light, ecology, geology, the solar system, and motion in physics. During the in-service program key concepts of each topic were identified as the focus for intended learning outcomes, and this identification of the topic's "big ideas" provided the starting point for discussion about possible writing types that could be used by students to re-represent understanding of these concepts. For example, a brochure to be displayed in a local bank was used as the basis for students to demonstrate their understanding of the problem of blue and green algae in the local water system. Brochures were also used as a basis for students to demonstrate their understandings of local geological features of the region (Hand & Prain, 1995). The topic of the properties of light was explored through students having to create a diagram solution to the problem of using convex and concave mirrors to enable sunlight to illuminate a mineshaft (Hand, Prain, & Scholes, 1997). Year 10 students were asked to create a captioned diagram to explain how they guided a spacecraft in space (Prain, Hand, & Kay, 1996). In addressing other topics students wrote narratives to show their knowledge of a process or a sequence of events, such as the life story of a chromosome or of blood cells. They also wrote travelogues and brochures on the solar system, as well as descriptions of a process, as in the topic of energy. They wrote guidelines and instructions, including survival manuals and clothing advertisements to demonstrate understandings of heat loss and retention in desert and arctic settings.

Some writing tasks entailed redrafting of first efforts and reworking initial concepts maps at the end of a topic. Students also produced concept and mind maps to show their initial understandings, to review and clarify topics, and to group ideas. They made posters to demonstrate their knowledge at the completion of topics. They assessed written work by other students to identify why one student's written script on a topic was more convincing than another student's. Students also wrote letters to peers, and textbook explanations for younger readers.

The implementation of varied types of writing led to the recognition that student writing could serve a diverse range of purposes, rather than only function as the means to assess student knowledge at the end of a topic. The teachers devised tasks with a broad diversity of purposes (see Figure 1), where the teachers made sure that these reasons for writing were made thoroughly explicit to the students.

By writing for different readers, students were expected to clarify their understanding of a topic by taking account of new readers' needs and background knowledge. For example, if students rewrote part of a textbook for younger readers they had to consider the knowledge base of this readership. Student wrote for a wide range of readerships, including themselves, other students in their class, younger students, the teacher, students at another school, parent groups, their own parents, consumers, politicians, or visitors to the classroom or school.

In general, students in the study had not written in diverse genres in science, and were only familiar with traditional writing tasks, such as note-taking, laboratory reports, and short-answer topic tests. A framework to guide teacher planning and implementation of writing tasks was devised, arising from the planning, trialing and evaluation of various kinds of writing tasks (Connor, Hand & Prain, 1994; Hand & Prain 1995; Prain, 1995). This framework was intended to represent the complex interconnectedness between the demands of different writing tasks and types, subject-topic-task interactions, and student learning outcomes. The framework identified five elements as constituting crucial dimensions to task demands as the basis for a matrix for learning through writing in science. These were: (a) writing types; (b) writing purposes; (c) audience or readership; (d) topic structure including concept clusters; and (e) method of text production, including how drafts are produced, both in terms of the technologies used as well as variations between individual and composite authorship processes. These five elements are conceptualized as analogous to interlocking keys within a combination lock. Writing tasks within the traditional science classroom are also based on the recognition of each of these elements, but task demands across these elements tend to be defined narrowly. Students typically write individually in a limited range of genres primarily for evaluative purposes, where the teacher is the main or only reader. By contrast, this framework recognizes that each key can represent a variety of task choices and demands (see Figure 2), and that there are multiple ways to structure effective combinations for enhanced student learning.

While particular combinations of choices within this matrix were effective for consolidating student learning through writing in specific topics, the framework was based on the assumption that diversification of students' purposes and kinds of writing, before, during and at the end of topics, enhanced learning outcomes generally. The framework's implications for teachers' pedagogical practices in implementing and assessing more diverse student writing-to-learn strategies were analyzed during the in-service program.

	Purpose for Writing
Start of Topic	Review students' current understandings Hypothesize (speculate on explanations) Explore first thoughts, preliminary reasoning Devise a plan of action
During	Clarify, refine knowledge, sort out understanding Reprocess ideas into own words Build outlines Structure an overview, classify information Argue ideas, take sides on an issue Revise understanding Reorganize and review knowledge Redraft to consolidate understanding Apply new knowledge to different context Consider alternatives Explain another writer's ideas Persuade others about a metaphor Test the validity of own or others' explanations Interpret data or ideas of others
Topic Completion	Demonstrate understanding Apply learning to new situation or problem Test learning Revise initial ideas Write notes as a resource Design a new experiment

Figure 1. Purposes for writing.

Issues Arising from the First Research Phase

Issues arising from the first research phase included both pedagogical and assessment concerns. The introduction of diversified writing types posed a range of challenges for teachers in terms of devising, implementing and assessing writing tasks. These challenges included designing effective matches between topic, genre, writing purpose, and method of text production. The teachers were also concerned about how to assess effectively the science concepts within the students' diverse language choices. The implementation of diversified writing practices also encouraged teachers to rethink the role of language as a medium for learning rather than as a neutral tool for assessment of student's concept knowledge. This initial exploration of ways to diversify writing types for learning science raised a range of issues that were reported in a case study of the teachers' perspectives (Prain & Hand, 1996a, 1996b). These included specific classroom strategies to increase opportunities for students to use writing to clarify and deepen their thinking, criteria for evaluating student writing, effective use of past student writing to clarify task demands for students, and strategies to introduce new topics.

Another ongoing concern was the question of the learning gains achieved by this approach over traditional uses of writing for learning in science. The small-scale qualitative studies in this research provided a rich picture of teacher and student perceptions and practices, but failed to quantify learning outcomes. The second phase of the research sought to investigate both implementation issues and quantification of the benefits of this approach to student learning.

THE SECOND RESEARCH PHASE: QUANTIFYING THE EFFECTS OF FACTORS ON WRITING TO LEARN SCIENCE

The second phase of our research, from 1997 onwards has attempted to focus on the identifying and implementing key contextual elements in using writing-for-learning strategies. There were two key ideas that we were concerned with in moving forward on useful designs for continued research. The first idea centered on the translation of the model into a realistic pedagogy that would enable successful implementation of writing-to-learn strategies. There was a need to explore more fully the teacher's role in supporting the learning processes associated with using writing. The second key idea, based on the work by Galbraith (1998), focused on the role of planning as a necessary component of success with writing. Planning involves much more than simply writing down some points or structure for completing the writing task. Rather the concept of planning involves the complete process of determining the information to be dealt with, negotiating the meaning of this information through interactions with others, the initial drafting of text, and the opportunity for redrafting of text.

Figure 2. Model of the five critical elements of a writing task.

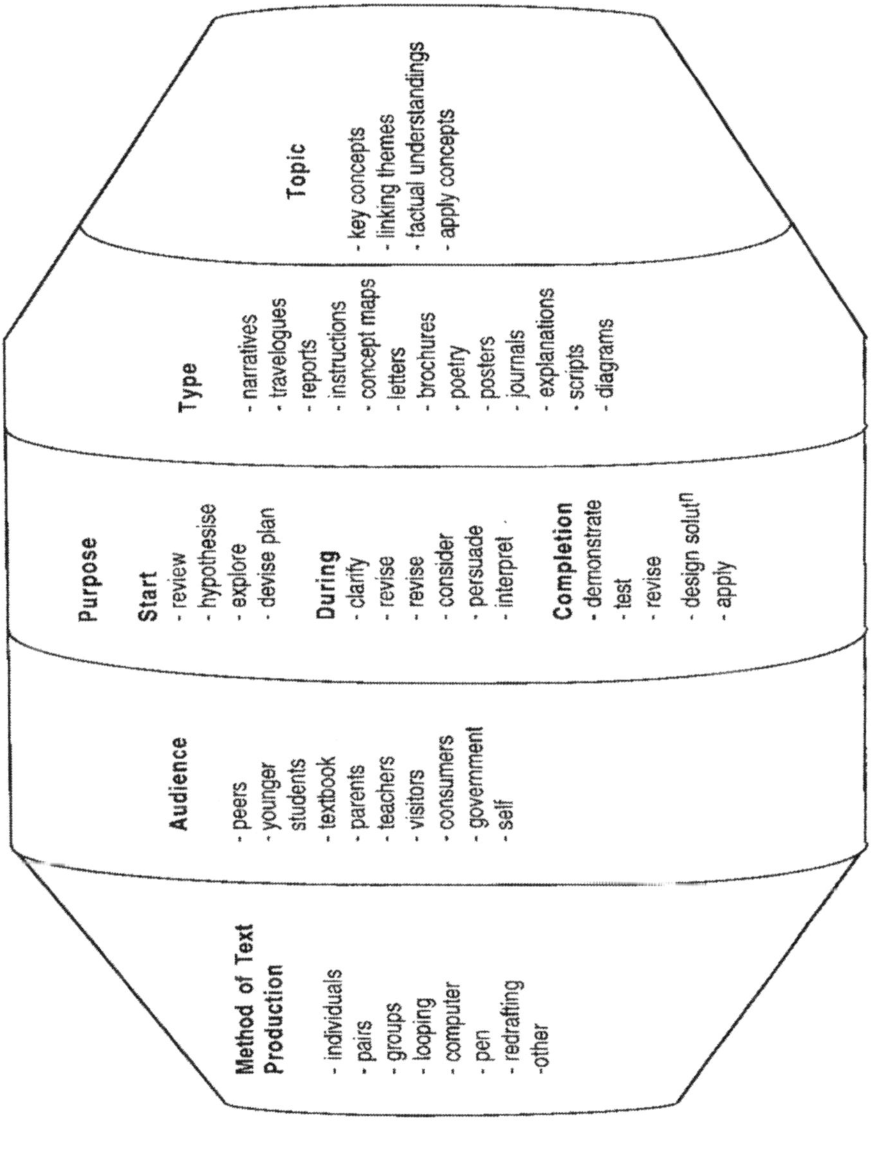

Study One

Our first study involved working with two teachers, one a male teacher with 27 years of experience and the other an inexperienced female teacher, on a unit on genetic engineering as part of a biotechnology subject. There were four classes of year 10 students with a total of seventy-three students involved. The study was centered on attempting to address two research questions:

1. Do students experiencing planning sequences prior to writing perform better on conceptual measures of content knowledge than students with delayed-planning experiences?
2. Is there a cumulative benefit to be gained from using multiple, non-traditional writing tasks?

The concept of planning in this study was structured to incorporate three elements: opportunities to negotiate meaning of science concepts, delineating the big ideas to be addressed within the text, and consideration of the audience for which the text was written. The teachers ensured that time was provided for both small group and whole group discussions to facilitate meaning negotiation. In the small groups, students were able to outline what they believed was essential to include in the writing task, while for the whole group component students presented their small group discussions for feedback from the entire class. It was at this point that the students were then given an opportunity to complete an initial draft of their text. The science teacher provided initial feedback on the science content, while the students' English teacher provided feedback on the language components of the text. The students were then provided with an opportunity to produce a final draft of their text.

The second element of the big ideas to be addressed was an essential element, because it made explicit for the students the guiding concepts of the unit under study. The two teachers decided that there were three important concepts for the students to understand the topic of genetic engineering. These three concepts were addressed throughout the teaching of the unit and were required components of the writing task. The last element of the audience is something that we believe is the crucial component that makes writing to learn such a valuable strategy. Instead of writing to the teacher, students were required to write a textbook explanation to a year 7 group as the first writing task (Task 1), and to the general public for the second writing task (Task 2). Both of these audiences were involved in providing feedback to the students on their final drafts.

To address the first research question, the four classes were initially divided into two groups – two classes were to have a planned experience from the start, while two classes were to have a delayed planned experience.

The delayed-planned group had a treatment that more closely reflected the teachers' traditional instruction; that of no scaffolding before writing. Therefore, the delayed-planned group wrote their initial drafts without pre-writing instruction, while the planned group had a planning experience before writing their initial drafts. The delayed-planned group had a planning experience after completing their initial text drafts and before completing their final drafts, while the planned group wrote their final drafts based on the teachers' feedback.

To address the second research question the two groups were further divided such that one class from each group was asked to complete another writing activity using the same method as for the first treatment. In this case the students were asked to write a newspaper article for the general public, while the other two groups were asked to complete chapter summaries and answer end of chapter questions. An editor from the local newspaper assisted the students in terms of understanding the requirements for an article to be published. Thus, at the end of the project the classes had been divided into four separate groupings: class 1 had a single planned writing experience, class 2 had two planned writing experiences, class 3 had a delayed planned experience, and class 4 had two delayed planned writing experiences.

To enable comparison across the groups, student performance prior to the study was collected and used as baseline data. The topics covered during this semester were related to the nature of science, chemistry, cell biology, and genetics, the last of which provided important background information for biotechnology, the unit of study in this project. The mean baseline score of each of the four classes on the genetics unit exam were used to identify each class as either a high group ($\underline{M} = 84.25$, $\underline{SD} = 10.48$, $\underline{n} = 16$; and $\underline{M} = 82.21$, $\underline{SD} = 7.38$, $\underline{n} = 19$) or a middle group ($\underline{M} = 79.22$, $\underline{SD} = 10.09$, $\underline{n} = 18$; and $\underline{M} = 80.75$, $\underline{SD} = 8.03$, $\underline{n} = 20$). However, using Analysis of Variance, the mean baseline scores were found not to significantly differ among or between the four class groups. In terms of trying to assess the effectiveness of planned writing as a learning tool the students were given two tests. After completing the first writing task, all students completed Exam 1, consisting of 12 recall questions and three higher order conceptual questions. Instruction for the unit ended after the students completed the second writing task. All students then completed Exam 2, consisting of five recall questions and another three higher order conceptual questions, which were worded differently from those on Exam 1, but targeted the same concepts. All students also completed Exam 3, the end of the semester final exam, which included three higher order conceptual questions related to the biotechnology unit.

Results comparing planning experiences indicated that students with planned experiences prior to writing the initial draft for the textbook explanation (Task 1), scored higher than students with delayed-planned experiences on four dependent variables, Task1- English score, Task1-

Science score, Exam 1- Total score, and Exam 1- Conceptual Question 3. For Task 1 English Score (Tk1E), significant effects were found for the covariant of base F (1, 63) = 18.58 and the main effect of the type of planning experience F (1, 63) = 5.74, $\underline{MS_e}$ = 4.199. Therefore, on the final drafts of the textbook explanation (Task 1), students in the planned group (PL) received higher English scores than students in the delayed-planned (DPL) group.

For the Task 1 Science Score (Tk1S), significant effects were found for the covariant of base F (1, 68) = 17.79, p<.0001 and the main effect of the type of planning experience, F (1, 68) = 4.03, p=.048, $\underline{MS_e}$ = 5.914, . On the final drafts of the textbook explanation (Task 1), students in the PL group received higher science scores than students in DPL group.

For the Exam 1 Total Score (Ex1T), there were significant effects found for the covariate of base F (1, 68) = 17.69, p<.0001, the main effect of the type of planning experience F (1, 68) = 4.35, and gender F (1, 68) = 8.67 $\underline{MS_e}$ = 24.401. Students in the PL group scored higher on Ex1T than students in the DPL group and males scored higher than females. Finally, for the Exam 1 Third Conceptual Question (Ex1Cq3), there were significant effects found for the covariate of base F (1, 68) = 12.44, the main effect of the type of planning experience F (1, 68) = 2.83 (p < .10), and gender F (1, 68) = 13.21, $\underline{MS_e}$ = 1.600. Students in the PL group scored higher on Ex1Cq3 than students in the DPL group and males scored higher than females.

Results comparing the number of writing experiences indicate that students who wrote twice (2WR) scored higher than students who wrote once (1WR) on six dependent variables, Exam 2 Conceptual Question 2, Exam 2 Conceptual Question 3, Exam 2 Conceptual Question Total, Exam 3 Conceptual Question 2, Exam 3 Conceptual Question 3, and Exam 3 Conceptual Question Total. After Task 2 (newspaper article) was completed, students who wrote twice (2WR) scored higher on more conceptual questions than students who wrote once (1WR). This difference persisted when total scores for conceptual questions were analyzed for both Exam 2 and Exam 3. Again, gender differences were found over both groups, males scoring higher than females on Exam 2 Conceptual Question 2 and Exam 2 Conceptual Question Total.

For Exam 2 Second Conceptual Question (Ex2Cq2), there were significant effects found for the covariant of base, F (1, 65) = 15.13, the main

Table 1. Scores (Adj. M SE) on task 1 (textbook explanation), exam 1 total and 3rd conceptual question (CQ) for students in the planned (PL) and delayed-planning (DPL) groups by gender.

Group	Textbook English Mark	Textbook Science Mark	Exam 1 total score	Exam 1 3rd CQ
PL				
M	10.01*	9.47*	17.94*	4.10
SD	0.34	0.39	0 .79	0 .20
n	37	39	39	39
DPL				
M	8.79*	8.30*	15.48*	3.59
SD	0 .38	0 .43	0 .87	0 .22
n	31	34	34	34
Males				
M	-	-	18.44**	4.39**
SD	-	-	0 .76	0.20
n	-	-	42	42
Females				
M	-	-	14.97**	3.29**
SD	-	-	0 .90	0 .23
n	-	-	31	31

Note. Missing data due to absences resulted in different n.
$* p < .05$. $**p < .01$.

effect of the number of writing experiences $\underline{F}$ (1, 65) = 4.41, and gender $\underline{F}$ (1, 65) = 5.43, $\underline{MS_e}$ = 2.710. Students in the 2WR group scored higher than students in the 1WR group and males scored higher than females. For Exam 2 Third Conceptual Question (Ex2Cq3), significant effects were found for the covariant of base $\underline{F}$ (1, 65) = 18.56, and the main effect of the number of writing experiences, $\underline{F}$ (1, 65) = 7.75, $\underline{MS_e}$ = 4.270. Students in the 2WR group scored higher than students in the 1WR group.

For Exam 2 Conceptual Question Total (Ex2CqT), there were significant effects found for the covariant of base $\underline{F}$ (1, 65) = 29.67, the main effect of the number of writing experiences $\underline{F}$ (1, 65) = 6.97, and gender $\underline{F}$ (1, 65) = 6.82, $\underline{MS_e}$ = 13.025. Students in the 2WR group scored higher than students in the 1WR group. Results from Exam 2 measures indicate that students in the 2WR group scored higher than students in the 1WR group on a greater number of conceptual questions on Exam 2. For Exam 3 Second Conceptual Question (Ex3Cq2), significant effects were found for the covariant of base $\underline{F}$ (1, 65) = 19.76 and the main effect of the number of writing experiences, $\underline{F}$ (1, 65) = 22.52, $\underline{MS_e}$ = 1.949. Students in the 2WR group scored higher than students in the 1WR group.

For Exam 3 Third Conceptual Question (Ex3Cq3), significant effects were found for the covariant of base F (1, 65) = 11.06 and the main effect of the number of writing experiences, F (1, 65) = 7.36, MS_e = 1.184. Students in the 2WR group scored higher than students in the 1WR group. Finally, for Exam 3 Conceptual Question Total (Ex3CqT), significant effects were found for the covariant of base F (1, 65) = 33.20 and the main effect of the number of writing experiences, F (1, 65) = 18.81, MS_e = 6.738. Students in the 2WR group scored higher than students in the 1WR group. Results from exam measures indicate that students in the 2WR group had a greater retention demonstrated by their ability to score higher on two conceptual questions on the delayed post-test, Exam 3.

The findings of this study are important because they provide very good evidence that planned writing experiences are beneficial. Students' performance on higher order conceptual questions were significantly better when they had opportunities for planning an initial draft, compared to students who had a delayed planned experience. Similarly, the opportunity for a second writing experience resulted in significantly greater understanding compared to only one writing experience. To confirm the outcome of the planning experience treatment, two additional studies were implemented.

Table 2. Conceptual question (CQ) scores (Adj. M SE) of students who wrote twice (2WR) and students who wrote once (1WR) by gender.

Group	Exam 1 2nd CQ	Exam 2 2nd CQ	Exam 2 3rd CQ	Exam 2 total CQ	Exam 3 2nd CQ	Exam 3 3rd CQ	Exam 3 total CQ
2WR							
M	2.40	6.50	8.01	17.24	3.47	7.19	13.21
SD	0.22	0.30	0.37	0.65	0.19	0.24	.45
n	35	33	33	33	35	35	35
1WR							
M	1.66	5.65	6.59	14.89	2.75	5.57	10.47
SD	0.21	0.28	0.35	0.60	0.19	0.24	0.46
n	38	37	37	37	35	35	35
Males							
M	-	6.54	-	17.22	-	-	-
SD	-	0.26	-	0.56	-	-	-
n	-	41	-	41	-	-	-
Females							
M	-	5.61	-	14.92	-	-	-
SD	-	0.31	-	0.67	-	-	-
n	-	29	-	29	-	-	-

Note. Different n are due to missing data resulting from absences.

Study Two

The focus of this study was on determining if writing to learn strategies would be useful in helping students construct understandings of stoichiometry. The study involved three Year 11 chemistry classes in a small rural Mid-western school that were involved in a three-week unit on stoichiometry. Building from the study above, this study incorporated the same rich context for planning for the treatment group, and extended the involvement of the audience in terms of providing feedback to authors. In the previous study, the students wrote for a real audience who evaluated the final product. To change this, we involved the intended audience after completion of the initial draft, and then on completion of the final draft. This enabled the student authors to be more attuned to their audience and to have the opportunity to address audience concerns as a function of the writing process. The research question for this study was: Do students who have extended involvement with writing to learn, including audience feedback, perform better on conceptual questions for stoichiometry than those without such an experience?

The classes were divided into two groups, with one class being a control in that they completed the normal activities that the teacher would have done without the writing treatment, and two classes that completed the writing treatment. The unit was design around three essential concepts of stoichiometry. At the end of the unit, the control class was asked to write a chapter summary and answer end of chapter questions from the textbook. The treatment classes were asked to write a business letter to Year 7 students explaining the concepts of stoichiometry. The planning activities used for the writing treatment reflected those from the first study with the one major alteration being that the initial draft of the letter was given to the Year 7 students for evaluation. That is, instead of waiting until the final draft of the writing to be completed before handing the product to be read by the intended audience, the audience was involved earlier. The Year 7 students were provided with a basic evaluation sheet asking them to comment on the whether they were able to understand the science ideas being discussed in the text, and if the text was well written. The Year 11 students were then provided with an opportunity to respond to the audience feedback in completing the redraft of the text.

The study was centered on a pre/post test design involving short answer questions and three higher order conceptual questions. The three higher order conceptual questions were classified as being the following. Question five was viewed as an extended recall question in that it asked students to do two part calculations in solving typical stoichiometry problems. Answering such questions was incorporated into the teaching strategies as part of the teaching unit, and thus the prediction was that both treatment and control groups would handle this question equally well. Question six was a transfer question in which students were required to transfer their knowledge of stoichiometry to a novel situation which had not been dealt with in class. Our prediction was that the treatment group would do better than the control group. Question seven was an analogy question requiring students to describe some of the processes involved in stoichiometry by using an analogy such as cooking. Our prediction was that the treatment group would do better on this question than the control group. However, we should note that some of the students in their writing described stoichiometry using a similar analogy and thus these results from this question have to be treated with caution.

T-test analysis was completed for the short answer questions (questions 1-4), questions 5, 6 and 7 and the total score. The results indicated that there was no statistically significant difference between the control and treatment groups in any of the categories. For total score, t (47) = 1.246, for the short answer questions 1-4, t (47) = 1.397, for question 5, t (47) = 0.527, question 6, t (47) = 0.757, and for question 7, t (47) = 1.368.

T-test analysis was completed for the short answer questions (questions 1-4), questions 5, 6 and 7, and for total score on the test completed by the students at the end of the teaching unit on stoichiometry. The results indicated that for the short answer questions 1-4 (t (50) = 1.544) and question 5 (t (50) = .408) there was no statistical significant difference between the treatment and control groups. For question 6 (t (50) = 2.296, p=0.026), question 7 (t (50) = 2.673, p=0.01), and total score (t (50) = 2.082, p=0.043) there was a statistically significant result.

These results support the outcomes from the first study by showing that there is a benefit from using writing to learn strategies for improving student understandings of science concepts. Importantly for us, the involvement of the audience in both editions of the draft was valuable in helping students construct understandings of difficult concepts related to stoichiometry. The third study reported here, followed from this study in that the concept of writing to a real audience was extended to developing a newspaper article for the local paper for students in a biology class.

Study Three

This study involved Year 10 students in a biology class who were studying a topic on molecular biology. The teacher of the class was the same experienced teacher from the first study who had agreed to continue with the research program. Two classes were involved in this study. The treatment group had 21 students and the control group had 22 students. The same rich planning experiences were provided for the students as in the previous two studies. In this study the students in the treatment group were asked to complete a newspaper report to the general public about molecular biology.

In order to make the experience real, the editor from the local newspaper was invited to speak to the class and outline the requirements for producing an article. The editor agreed to read the initial drafts of the students and provide feedback so that the students could amend the final drafts. The students were also told that there was a definite possibility of having their work published if the final product was good. The six-week long molecular biology unit centered on three essential concepts, and these were to be the focus of the newspaper article. The research question we asked was: Do students who are experiencing a realistic context and audience for writing perform better on test questions of conceptual understanding in molecular biology than those who do not?

A similar research design to study two above was implemented in this study, featuring a pre-post structure. To ensure that equal time on task was given to each group, the control group of students was asked to complete end of chapter questions and a chapter summary during the time allocated for the writing treatment. The treatment students were asked to write a newspaper paper article for the general public in which they asked to provide an explanation of the three concepts framing the topic. The instructions for the newspaper article were provided by the editor of the local newspaper. He agreed to come in during one class period and outline the framework of a newspaper article. During the period he provided guidance to students where needed. Students were given two class periods plus a one scheduled homework activity to complete the task. The editor and the teacher provided feedback on the students' written products prior to the test. Three higher order conceptual questions were added to the multiple choice/short answer test format, for example, one question was describe the structure of carbon and explain how its structure relates to its function. Student scores on Iowa Test of Educational Development were used as a co-variate in order to test for group difference. No significant difference was detected.

Results indicated that the treatment students performed better than the control group on two of three higher order conceptual questions. Significant difference between the treatment and control groups was noted for higher order conceptual question 2 ($M = 1.55$, $SD = 1.30$; $M = 0.27$, $SD = 0.63$, respectively) $t(42) = -4.13$, $p = 0.000$, and higher order conceptual question 3 ($M = 1.72$, $SD = 0.55$; $M = 1.27$, $SD = 0.83$, respectively) $t(42) = -2.15$, $p = 0.038$. These differences persisted on the total conceptual question score of all three higher order questions with the treatment group ($M = 4.91$, $SD = 2.43$) performing better than the control group ($M = 2.64$, $SD = 1.76$), $t(42) = -3.55$, $p = .001$.

Again, these results confirm the value of using writing as a learning tool. Those students who wrote to a different readership from the teacher performed better on higher order questions than students who completed end of chapter questions and wrote a chapter summary. By having to write and explain what they understood within the topic, the students were having to construct a richer understanding of the topic themselves. This understanding was reflected in their performance on questions that required them to translate their understandings to new contexts.

FUTURE RESEARCH AGENDA

Our research agenda will continue to focus on implementation and assessment issues, including the following topics: (a) analysis of effective topic-genre-writing purpose choices in writing for learning; (b) identification of student cognitive processes when they engage with learning from writing tasks; (c) analysis of the learning effects of diversified writing tasks on different classroom populations; and (d) identification of the role of writing for learning within new larger multi-modal accounts of science practices. While there is emerging evidence of the value of students writing in diverse ways to learn science, there is still a range of issues that require further investigation to identify conditions that maximize learning gains from this approach.

CHAPTER 6

USING A SCIENCE WRITING HEURISTIC TO PROMOTE LEARNING FROM LABORATORY

Carolyn S. Wallace & Brian Hand

In 1997, Hand and Wallace (formerly Keys) created a tool for enhancing learning from laboratory activities through writing to learn. The tool, called the Science Writing Heuristic (SWH, see Figure 1), consists of both activities and metacognitive support to promote reasoning about laboratory data and concepts. Similar to Gowin's Vee heuristic, the SWH provides learners a template for thinking, doing, and writing. In addition, the SWH provides teachers a template of suggested activities to support the hands-on portion of labs. As a whole, the activities and metacognitive scaffolds are aimed at generating authentic meaning making on the part of the learner. The negotiation of meaning occurs across multiple formats for talk and writing. The teacher template includes eight possible activities surrounding a laboratory: (a) exploring pre-instruction ideas through concept mapping; (b) engaging in pre-laboratory writing activities, such as brainstorming, free writing or posing questions; (c) participation in the laboratory activity; (d) writing personal meanings for the laboratory activity, such as journal writing; (e) sharing personal interpretations of the data with peers in small groups; (f) comparing laboratory data with relevant ideas in printed resources, such as the textbook; (g) engaging in individual reflection and composing a product for public viewing, such as a laboratory report, newspaper article or PowerPoint presentation; and (h) engaging in post instruction concept mapping.

We conceptualized the SWH as a bridge between informal, expressive writing modes that foster personally constructed science understandings, with more formal, public modes that focus on canonical forms of reasoning in science. Thus, the heuristic scaffolds learners in both understanding their own lab activity and connecting it to other science ideas. The template for student thinking (see Figure 2) prompts the learner to generate questions, claims, and evidence for claims. It also prompts them to compare their laboratory findings with others, including their peers and information in the textbook or Internet. Finally, the template for student thinking prompts the learner to reflect on how her own ideas have changed during the experience of the laboratory activity. The SWH may be used in its generic form (as shown in Figures 1 and 2) or it may be tailored to specific activity contexts through reformulation of the scaffolds. We have tried both approaches in our research on the SWH in classrooms. One way to think of the SWH is as an alternative format for laboratory reports. Instead of responding to the four traditional sections, purpose, methods, results and conclusions, students respond to prompts eliciting knowledge claims, evidence, description of data and observations, methods, and reflection on changes to their own thinking. We have been conducting both quantitative and qualitative research studies on the use of the SWH in actual classrooms from seventh grade life science to college freshman chemistry over the past five years. In this chapter, we will describe the theoretical framework that underlies the development of the SWH and a summary of the research findings.

THEORETICAL FRAMEWORK

Recently, scholars of scientific English have argued for *semiotic* approaches to understanding the use of language in science (Berkenkotter & Huckin, 1995; Halliday & Martin, 1993). The study of semiotics asserts that language must be considered with respect to its function of meaning making in the culture in which it is used. The characteristics of scientific English, such as that found in science journal articles, have evolved over centuries to serve specific needs of the scientist in communication. The unique grammatical forms, compact nature, and detached voice used in science writing have arisen due to the needs of the scientific community. As a by product of this process, the compressed language of science *makes it seem as though* reporting is a recapitulation of knowledge already known to the writer, when in fact, it is knowledge that has been negotiated and renegotiated many times previously through informal writing, peer group discussion, reading, and thinking. Therefore, the traditional four-part laboratory report with sections for purposes, methods, results, and conclusions tends to obscure rather teach science learners how knowledge is developed and communicated in science. The traditional scientific report tends to "leave implicit precisely the experiential meanings that they [learners] most depend on for its [scientific understanding] construction," (Halliday & Martin, 1993, p. 68). Since meanings are implicit, the traditional laboratory report may be a poor tool for the construction of science knowledge.

Writing in science classes may be most useful if it reflects a pedagogical orientation, as well as a disciplinary orientation; that is, the writing genres used in school science should reflect some of the characteristics of scientists' writing, but should also be shaped as pedagogical tools to foster students in "unpacking" scientific meaning. We conceived of the Science Writing Heuristic, then, as a vehicle to illuminate the construction of understanding from lab. The SWH promotes both cognition about the laboratory and metacognition, or how the learner has to come to know what he or she does know. It is a vehicle for science learning and therefore exemplifies writing to learn principles. Because the SWH capitalizes on canonical forms of scientific thinking, such as the coordination between claim and evidence, it has the potential to build nature of science understandings, as well as conceptual understandings.

The SWH differs from a conventional laboratory report in some important ways. First, writing activities are used before, during, and after laboratory work, communicating to the learner that writing is used to generate knowledge throughout the investigative procedure. Second, the SWH emphasizes the collaborative nature of scientific work, as learners are purposefully engaged in negotiating meanings with their peers. Third, the student template encourages learners to make connections between questions, observations, data, claims, and evidence. When students state a claim for their investigation, they are asked to describe a pattern, make a generalization, state a relationship, or construct an explanation. They are then required to provide data and evidence to support that claim, and examine how their claim compares with print-based readings. Fourth, learners are engaged in reflecting on their own knowledge growth throughout the activity, enhancing metacognition.

An additional theoretical idea is that the nature of writing laboratory reports is highly intertwined with the nature of the investigative activity itself. Traditional reports are a fairly good match for cookbook labs, which generally do not allow for authentic questioning and data interpretation. When the procedures are the same for everyone, the data is similar, and the claims match what is expected to happen, then reporting results and conclusions becomes a rote process. In authentic inquiry, the students may have license to pursue their own questions, design their own procedures, or make sense of data patterns that were not expected as a demonstration of theory already known. When these events occur, the SWH helps provide scaffolding for authentic thinking about questions, evidence, and claims.

In addition to the studies described in detail below, some of our colleagues have been researching the use of the SWH in college level freshman chemistry. Rudd, Greenbowe, Hand, and Legg (2001) found that general chemistry students in an intervention using the SWH for their laboratory reports demonstrated (a) significantly higher median scores for written explanations on an equilibrium lecture exam question; (b) significantly higher success rates on an equation writing question on the lecture exam; (c) no statistical difference on a lab practical question, although there was a slight improvement for students using the SWH. In addition, the researchers conducted a survey on student attitudes about the SWH. The survey indicated that the vast majority of students preferred the SWH over a traditional report because they learned more, it took them less time to complete, and they felt it was less busy work.

Figure 1. Student and teacher templates for the Science Writing Heuristic

Teacher Template

1. Exploration of pre-instruction understanding through individual or group concept mapping.
2. Pre-laboratory activities, including informal writing, making observations, brainstorming, and posing questions.
3. Participation in laboratory activity.
4. Negotiation phase I- writing personal meanings for laboratory activity (For example, writing journals).
5. Negotiation phase II- sharing and comparing data interpretations in small groups (For example, making a group chart).
6. Negotiation phase III-comparing science ideas to textbooks or other printed resources (For example, writing group notes in response to focus questions).
7. Negotiation phase IV- individual reflection and writing (For example, writing a report or textbook explanation).
8. Exploration of post instruction understanding through concept mapping.

Student Template

1. Beginning Ideas --What are my questions?
2. Tests-- What did I do?
3. Observations-- What did I see?
4. Claims -- What can I claim?
5. Evidence-- How do I know? Why am I making these claims?
6. Reading -- How do my ideas compare with other ideas?
7. Reflection – How have my ideas changed?

METHODS

Context of the Studies

The Ames Middle School studies (Hand, Wallace, & Yang, in press; Wallace, in press) took place during 1999-2000 and 2000-2001. The studies were located in a suburban middle school in a mid-sized mid-western town in the United States. Both studies involved seventh grade life science classes. All classes in both Ames studies were taught by one teacher, who was part of our collaborative research group for writing to learn. At the time of the studies, he had approximately five years teaching experience and was working on his master's degree in science education. He is a white male who had developed an extensive background on science literacy through his interaction with the research team and graduate courses. The children in this school were primarily White, with only a small percent of Asian, Black, and Latino students. During the first Ames study, twelve seventh grade children were interviewed, and during the second study, ten children were interviewed. All these children had participated in writing to learn tasks including some combination of SWH report writing, writing a letter to a parent, writing a textbook explanation for a peer seventh grader, or writing a newspaper article. The participants were selected by the teacher and/or classroom researcher. They were purposefully selected according to the following criteria: balanced gender representation; not inhibited to talk to the researcher; balanced representation of low, middle, and high achieving students according to class grades and teacher perception.

The Ballard High School study (Tsoi & Keys, 2001) took place during the 1999-2000 school year in a tenth grade biology class in a suburban high school. The setting was the same mid-western town in which the Ames studies were conducted. The topics under study were cells and molecular biology. All students in the study were taught by a veteran, White male teacher with approximately 28 years of teaching experience. This teacher was also involved in graduate studies at a local university and interacted frequently with the writing to learn research group. Students in the Ballard High School were also primarily White. All students who were interviewed had participated in writing either SWH reports, or textbook explanations for seventh graders or both. Twelve students were interviewed based on the same criteria as above including, balanced gender representation, balanced science achievement, and willingness to talk to the researcher.

The Victoria study (Hand, Prain, & Wallace, 2002) took place during the 1997-1998 school year in a rural secondary college located in the state of Victoria, Australia. All students in the study were taught by the same teacher, a White male with fourteen years teaching experience at the time of the study. The teacher met weekly with the researchers to discuss ongoing progress of the study. The students at this school were primarily White.

Students who were interviewed in this study included twelve ninth graders who had recently completed a unit on the physics of light. All students who were interviewed had completed SWH reports and a letter to a peer student in the U. S. explaining what they had learned about reflection and refraction. They had also completed writing on an essay test question to create an analogy. The students were purposefully selected, based on a mix of gender and science achievement as determined by the teacher.

The South Hall Water Quality study (Keys, Hand, Prain & Collins, 1999) and the South Hall Erosion study (Keys, 1999) took place in a rural setting in a small town in the southeastern United States. Students in both studies were taught by the same teacher, a White, female with approximately 25 years of teaching experience. The teacher was also a graduate student at a nearby university and interacted with members of the writing research group. For the water quality study, interviewees were selected from two classes of eighth grade earth science students, representing both average and high achieving students. These students had participated in a water quality curriculum to investigate a small stream behind their school. The SWH was used in conjunction with the water quality curriculum. There were 16 target students in four teams of four in the water quality study, based on volunteerism and balanced gender. Data reported here come from group interviews of four students each in their working teams, or from written documents in the form of SWH reports. In the South Hall erosion study, again 16 students were selected for study in a clinical setting. Some of these students overlapped with the target students in the water quality study, but not all. Data reported here comes from tape recorded think-aloud protocols or written reports. The topic of this study was evidence of erosion on hills surrounding tennis courts on the school grounds.

Methods for the Quantitative Studies

In the Ames middle school 1999-2000 study, the Ballard High School 1999-2000 study and the Victoria study, a quantitative design was used to determine if there was an improvement in students' conceptual understandings as a result of using the SWH. The research was based on asking students to complete higher order conceptual questions on an end of unit test. Each of the tests were constructed in conjunction with each of the participating teachers and consisted of two parts. The first part was based on multiple choice or short answer questions, for example, fill in the blank or answers that only required a single sentence. All the teachers involved had used these types of questions previously on their tests. The second part of the test involved the addition of higher order questions that required extended answers from the students. The intention of these questions was to require the students to engage more extensively with their conceptual understandings of the topics under review than was required by the multiple choice or short answer questions.

All the studies involved using control and treatment groups at each site, with a single teacher at each site instructing both groups of students. The Victoria study involved a post-test only design, with the Ames and Ballard study involving a pre-post test design with the same test used as both the pre and posttest. The Ames and Ballard studies involved the addition of three higher order questions, while the Victoria study involved the addition of one higher order question.

For each of the studies, analysis of covariance was used as the means of analysis. Scores on previous science test were collected and used as a co-variate in order to account for any variance between groups of students at any one site. The results were analyzed in order to determine if there was a statistically significant difference on the following: multiple choice/short answer total, scores on individual higher order questions, higher order questions total score, and total score on test.

In the Victoria study the total number of students involved was 52, with the control group being 25 students and the treatment group being 27 students. Scores on two previous tests were used as a co-variate. The students completed five laboratory activities before completing their letter activity as the summary report. The additional question added to the post–test asked students to describe and explain what would happen to the laser beam and the bullet as a laser-targeted bullet entered a body of water.

In the Ames Middle School study there were 88 students involved. The students were divided into three groups. The first was the control group (N=18), who implemented normal laboratory activities and on completion wrote a summary report to the teacher, the second group (N=35) implemented the SWH activities and wrote a summary report to the teacher, while the third group (N=35) completed the SWH activities but wrote a textbook explanation for younger students as their summary report. All students completed three laboratory activities, with the two treatment groups using the SWH for each. Scores on the Stanford Diagnostic Reading Test was used as the co-variate. Three higher order questions were added to the teacher's end of unit test. The following is one of these questions: As a scientist in 1842, you have determined that all living things have different levels of organization. For instance, you think that all living organisms have at least a cellular structure of some kind. Many of your colleagues disagree with this idea. Develop an argument with evidence to support your views.

In the Ballard school study there were 39 students involved with the control group having 18 students and the treatment group having 21 students. The students' scores on the Iowa Test of Educational Development were used as a co-variate. Each student completed three laboratory activities before undertaking the summary writing activity, which for the treatment group was a textbook explanation for year 7 students. Three higher order questions were added to the teachers' end of unit test. The following is one of these: We can use analogies to describe things. So if Ames (a city) was a cell, what in Ames would be the mitochondria? Explain.

Methods for the Qualitative Studies

In three of the studies, students were interviewed individually by a member of the research team, in a quiet area of the school such as the media center. Interviews lasted from 20-45 minutes and were audio recorded and transcribed in full. In the Ames, Ballard and Victoria studies, we asked the children questions about their experiences using the SWH, for example, which activities of the SWH they found most helpful, what they did in their groups, how they felt about writing, what were their views of claims and evidence, and how their learning compared with that in previous science classes (see Appendix A for interview questions).

In the South Hall Water Quality study, students were interviewed in teams of four, which comprised their working lab teams. Each team met with the researcher in a small conference room adjacent to the media center. The researcher presented a question to the entire team and student volunteers from the team answered the question. Occasionally, the researcher would probe a student if s(he) had not been contributing much by saying "[student's name], what do you think?" These group interviews were audiotaped and transcribed in full. In the South Hall Water Quality Study, we were especially interested in students views on the nature of science after participating in SWH activities. The interview questions for this study are found in Appendix B.

In the South Hall Erosion study, think-aloud protocols were used rather than interview questions. The research was conducted with the 16 participating students interacting in teams of four to engage in a laboratory activity on erosion, using the teacher and student protocols for the SWH (The think-aloud protocol transcript may be found in Appendix C). Each participant took part in a practice session for thinking aloud, by saying everything that came to mind while writing about a favorite person, pet, or hobby. During the data collection period, students wrote an essay on the topic of erosion on their school grounds while thinking aloud. The essay was structured with guideline writing prompts shown in Appendix D. The students were reminded to keep talking continuously during the process. All think-aloud protocols were audiotaped and transcribed in full.

Interview and think-aloud protocol data were analyzed through established qualitative coding methodologies, particularly, descriptive and interpretive coding (Miles & Huberman, 1994). Patterns in codes were noted. These patterns were explored by testing hypotheses against the data itself. Major patterns that held up under scrutiny of examining the data sets were used to construct assertions. For this volume, a secondary analysis of common patterns across the various studies was implemented. An exception was made when writing assertion # 1 below. It is based on data from only one study, the South Hall erosion study. Written document data was coded used to support assertions in a manner similar to verbalized data.

RESULTS

Findings from the Quantitative Studies

The findings from the three quantitative studies are presented individually and then discussed. Importantly, the data for performance on the higher order conceptual questions is provided for comparison of the value of using the SWH across different groupings and topics.

For the Victoria study, in which only one higher order question was asked, there was a statistically significant difference between treatment and control on the analogy question using t-test analysis. For the multiple choice/short answer questions there was no significant difference (t=2.87, p .1), while for the analogy question there was a significant difference (t=4.98, p 0.001). After removing the effects of the previous science scores using ANCOVA, the SWH group still performed significantly better on the higher order question than the control group (F(1,41)=15.85, p=.001). The scores on the multiple choice/short answer questions indicated no statistical significance (F(1,41)=4.85, p=.032).

For the Ames study, analysis of the Stanford Diagnostic Reading Test indicated that there was no significance between the control and the two treatment groups prior to the study. Post-test scores on the conceptual questions and the multiple-choice question were analyzed using ANCOVA. The ANCOVA on the post test conceptual question scores revealed only significant main effect of the groups, F (2, 86)=13.63, p<0.00, MSE=29.60. The pair-wise test results reveal that the SWH textbook explanation group outperformed the SWH teacher report group and the control group. To investigate the group differences on each conceptual question, the follow up analyses were performed for each conceptual question item. For the first conceptual question, there was significant main effects on groups (F (2, 86)=8.21, p<0.00, MSE=43.44). The students in the SWH teacher report group and the SWH textbook explanation group achieved better scores over the students in the control group on conceptual question one. The SWH textbook explanation group of students performed better over the other two groups on conceptual question two, (F (2, 86)=9.78, p<0.00, MSE=5.48) and conceptual question three, (F (2, 86)=7.11, p<0.00, MSE=6.80). For the multiple-choice questions the ANCOVA found a significant main effect of the groups (F (2, 86)=6.03, $p<0.00$, MSE=16.16) The pair-wise comparison results showed that the SWH textbook explanation group and the SWH teacher report group outperformed the control group. There was no significant difference between the SWH textbook explanation group and the SWH teacher report group on the multiple-choice questions.

For the Ballard study, there was no significance difference between the treatment and control groups prior to the study on analysis of the Iowa Test of Educational Development. Analysis of the conceptual questions provided different patterns of results obtained in the previous studies. The results from t-tests analysis indicated that students studying cells in the control group, (M = 2.22, SD = 1.48) performed better on conceptual question one than students in SWH treatment group (M = 1.05, SD = 1.40), t (37) = 2.55, p = .015. However, these findings were reversed for conceptual question three, where the SWH group (M = 3.48, SD = 0.60) performed better than students in the control group (M = 2.83, SD = 0.51), t (37) = -3.55, p = 0.001. There was no significant difference between the groups for conceptual question two.

The trend of the results obtained across these studies was for the SWH treatment students to outperform the control students when answering higher order conceptual questions on end of unit test. The difference between the Ames middle school study and the Ballard study is interesting because the topic in both studies was cells. Thus, while the treatment in the middle school study produced significant statistical results, the results for the Ballard study were only positive for one of the three questions. Possible explanations may be related to familiarity with the topic as high school students have previously studied cells, poor question design, that is, the questions did not properly address the conceptual areas covered in the unit, differences between the teachers or other such factors. Overall the results indicate that the SWH does hold potential for improving students' understandings of conceptual knowledge of the topics studied and should be implemented more widely.

Findings from the Qualitative Studies

Our research studies on the use of the SWH in classrooms have, in general, shown that the heuristic does indeed stimulate metacognition and reflection with students (Please see also, Chapter 7 in this volume for children's views of how writing fosters learning). However, there are some important caveats; some classroom situations provide a better context for metacognition and reflection than others and some students are more inclined to take advantage of opportunities for metacognitive thought.

Assertion 1: Evidence indicates that the act of writing an SWH report can stimulate metacognitive self-talk for some writers in the form of searching for hypotheses, evidence, claims, or meanings.

The findings from the South Hall erosion study indicated how students think during the actual writing of an SWH report that was embedded within a series of SWH activities. The aim of the study was to investigate whether students actually engaged in the cognitive process described by Klein (1999) as backward search. Bereiter and Scardamalia (1987) originally hypothesized backward search as the process taking place when writers form a writing goal, and then search for information to accomplish that goal. Thus, Keys investigated whether eighth grade writers would engage in backward search during the writing of a report on erosion, using a think-aloud protocol technique. The relatively young writers were scaffolded with an entire SWH activity series, including small group discussion on their authentic data, and supplied with overall writing goals for their report.
Five of the 16 participants exhibited a think-aloud protocol with over 90% verbatim dictation. The evidence indicates that these five students did not engage in backward search during report writing, because almost all of their think-aloud talk was recorded directly into the report without reflection or transformation. An analysis of their written reports, however, showed that the lack of backward search did not necessarily result in weak report writing. Two of these five students wrote reports rated as strong, on the criteria of including rich hypotheses, claims, and evidence. For example, Lisa, did not deliberate or search for information during the act of report writing, but produced the following excerpt:

> *The grass is worn down where people have walked, water has run down the hill, and where gravity and wind have moved the soil. The erosion is bad enough that rocks once under the surface have been exposed and gullies have formed. I measured the gullies. They seemed to vary in their depth due to width. The more narrow gullies were deeper and had more rocks exposed. This is probably because of the concentration of the water and how hard it must have been raining at the time. The wider gullies seem to be more shallow because the water spread out over a larger area (Lisa erosion report, June, 1998).*

Although there is no evidence that Lisa used backward search mechanisms, she readily poses probable explanations for patterns she saw in her gully measurements. She hypothesizes that the narrow gullies were deeper because of a deeper concentration of water and the wide gullies were shallower due to the spreading out of water. She uses her gully measurements and observations, such as rocks showing, to support her claims of erosion.

In contrast, 11 of the 16 participants in the study did verbalize outside of words that were being recorded into their report verbatim. Additional think- aloud talk ranged from 20-70% of the total. Some of this talk included rehearsing language choice, which was then repeated during dictation, making general claims about erosion, or engaging in rhetorical planning. Another portion of the talk demonstrated that some students were indeed engaged in backward search mechanisms. Coding indicated that these students were carrying out the following metacognitive and cognitive processes during writing: (a) evaluating hypotheses and data as evidence; and (b) and generating meanings for patterns in data, hypotheses and evidence. These thinking processes are exhibited in the think-aloud protocol generated by Justin, below. Backward search is indicated as he moves from the recording process back to the think-aloud process, for example, after he writes, "I also tested the depth of the trenches." Italics indicate the portions that eventually were written in the report.

> There seems to be gullies where the water runs down (generating hypothesis). Okay is what I observed the number of trenches and dead grass to support this theory (evaluating evidence for hypothesis)? I observed trenches where water has killed the grass (generating evidence). [dictates as he writes,] *"I observed that there were trenches where the water flooded the grass, killed the grass, and washed away parts of the soil. I also tested the depth of the trenches*. [Pauses writing]. the trenches, um, the depth showed the depth of rain due to the amount of water and the amount of sediment in the water that was carried over (generating hypotheses). The depth test showed (generating evidence) there was a difference due to the speed of the and the flow of the water (generating hypothesis), hmm, yeah (evaluating hypothesis). (Justin, think-aloud protocol, June, 1999).

Justin's data, as well as that of several other students in the study, indicates that some students engage in metacognition and reflection during the act of writing itself. Therefore, we believe the SWH can play a role in promoting metacognition about questions, claims, evidence, and hypotheses. In the study reported above, students were asked to make a claim about erosion on their school grounds, support their claims with evidence, and pose possible causes of erosion. Some students, like Justin, thought aloud about claims, evidence, and explanations, thus making thinking during SWH report writing visible. We still do not know much about the interaction of the SWH with the thinking of students like Lisa, who sit down and generate text rapidly without reflection. More research is needed to probe metacognitive thought during other portions of SWH activities or without the explicit SWH probes.

Assertion 2: Evidence indicates that students using the SWH have been engaged in reflection on and modification of science ideas.

Some of our studies, most notably the South Hall water quality study, the two Ames studies, and the Ballard study, have provided evidence that using the SWH activities does impact students' science conceptions. Qualitative data (in addition to the quantitative data indicated above) indicate not only that students have changed their science conceptions, but they have been metacognitive about the reasons and justification for that change.

In the Ballard study, tenth grade biology students used the SWH student template as a format to report on several fairly traditional biology labs. For example, the classroom teacher had the students observe cheek, onion, and pond water cells to help characterize, compare, and contrast features of living cells. The students and the teacher together during class discussion formulated simple questions to guide the investigations, for example, "How do a plant and animal cells differ?" and "Do cells move?" A qualitative analysis of writing on the student templates, primarily under the probes for claims, evidence, and reflection indicated that a substantial number of students changed some of their basic conceptions about cells towards more scientific views. By promoting the children's authentic ideas, the SWH seems to be effective for unpacking core misconceptions.

Coding of the template reports for the "cheek" lab indicated that of several students changed their initial idea that cheek cells would individually be red or pink to a new idea that they are clear or transparent. These students made claims that cells are clear, not pink or red, cited evidence that the cells became more visible after adding a dye, and reported the change in their conception of cell color in the reflection portion of the template. Similarly, surrounding these cell labs, several students mentioned a change in their idea that cheek cells could move independently, and several discussed their new conceptions of the differences between plant and animal cells. The following excerpts are quoted from Ballard students' cell labs template reports.

> *[From claims section] Cheek cells are animal cells because they don't have a cell wall. The cells were clear. They didn't move, and they didn't have a cell wall . . . [From reflection section] My ideas have changed about the ability of cells to move. I thought that cells could move by using an extra extremity. All of my other ideas were basically the same as before. Except my idea about the color that they would be to a degree. I thought they would be clear or pink. They appeared clear to me, but I think that if you would cluster them together, they would be pink in color (boy, Ballard study).*

> *[From evidence section] I don't think it [the onion cell] moves, because I didn't see it move. The cells all stayed in the same position the whole*

time I was looking at them. The cells are all in one big cluster. There wasn't a cell sticking out on it's own. Each one was connected to another and when you would blow on them, it would move as one, not as individuals. When compared to my cheek cells I could tell a huge difference. Cheek cells were by themselves; onion cells are clustered together. Onion cells could be seen without dye, cheek cells could not. Onion cells have a cell wall; cheek cells do not. Cheek cells are more round or oval like; onion cells were almost rectangles . . . [From reflection section] I thought that the two would be more alike. I didn't think I would find that great of a difference between them. I guess I think of plant cells as more of a cluster now. I thought that it would look just like the cheek cell or at least close. But they were completely different except for both having a nucleus (girl, Ballard study).

While these excerpts demonstrate some incomplete understanding of plant and animal cells, they do indicate authentic reflection on the laboratory. These students have constructed their own understandings of cells, rather than relying on rote information from the text. The templates provide the teacher with a useful tool for conducting further class discussion on more in-depth concepts related to cells, such as the causes of animal tissue looking red, or the differences in epithelial structures of plants and animals.

We also observed substantial changes to children's ideas about water quality during the South Hall Water Quality study. In this class, eighth grade earth science students conducted a 12-week study of the quality of water in a small stream behind their school. The SWH activities were used as a scaffold for promoting enhanced understandings of a water quality curriculum. The excerpts below indicated that students changed their ideas about the water quality of the stream, via the consideration of the evidence.

My opinion is that on our first trip to the stream, the majority of us students believed that the stream would be very polluted. But we really didn't have any strong evidence to say that our statement was true (girl, South Hall Water Quality study).

We took some tests on the stream that gave me a whole new perspective of the stream (girl, South Hall Water Quality Study).

I know it is good [pH] because we took the pH test of it and we determine [sic] that it was 7.5. By knowing the pH of it, we know that many organisms can live in it (boy, South Hall Water Quality study).

In this study, many of the students had originally believed, as stated in the first excerpt, that the stream would be very polluted or have a highly acidic pH (interview data from same study). Students came to change their ideas about the stream through performing scientific tests and gaining evidence. When they did readings about water quality factors, such as, healthy pH and dissolved oxygen levels in bodies of fresh water, reflection on their data led to new conclusions.

One of the goals of the SWH has been to promote increased understanding of the nature of science, as an important element of scientific literacy. By creating a bridge between everyday discourse and scientific discourse, we hoped to foster in students a sense of authentic ownership for scientific ideas. Prominent among the scaffolds of the student template are the prompts to formulate questions, make claims, and state evidence for claims. By working through these thought processes, students may experience first hand the heart of canonical science, reasoning from observations to evidence. Although some teachers provide their students with questions during the SWH, other teachers in our studies have left the question open for the students or have given students a choice. All teachers in our studies, have made the probes for claims and evidence a central feature of the SWH laboratory experience.

> Assertion 3: The SWH has been efficacious for promoting the understanding of questions, claims and evidence in science.

Interview and written report data with students from the Victoria study, the South Hall studies, the Ballard study, and the Ames studies indicate that the SWH has the potential to develop sound and sophisticated understandings of the nature of a scientific claim, how evidence is used to support a claim, and how a claim must correspond to a question.

First, students using the SWH for at least a few activities, develop an understanding of scientific claim as a statement based on inferences from and connected with the laboratory data. In the excerpts below, students talk about how they construct a knowledge claim. In the first interview quote, the participant talks about "seeing what connections are available" between the question and the data, as well as referring to gaining ideas from class discussion. In the second quote, the participant describes the process of constructing a claim in a group. The third participant describes a claim as an inference from observations, or "what you see."

> Basically, we normally had done an experiment before then, so I used and I (inaudible) preliminary data. So from looking at that just seeing what connections are available and what looks like a probable answer. And also from . . . I used what we had been talking about in class discussion too- where we discussed possible claims in class and to figure out which of those make the most sense and seems the most accurate with the data that we've gathered (boy, Ames study II).

> What my group does is compare all of our ideas, like, for instance, somebody says it first and then it goes on and then like somebody has an idea we compare and put it all in a group claim (girl, Ames study I).

> You saw something and you made a claim about it. So, like the example used in class was that you see Mr. S's room has science stuff, so the claim is Mr. S likes biology or science. You just make a claim by what you see (girl, Ballard study).

Second, when students use the SWH, they learn that the evidence is used to support the claim and there must be coordination between claim and evidence. They also come to distinguish between claim and evidence. The quotes below illustrate these ideas. A seventh grade boy from the Ames II study in the first quote indicates the distinction between claim as "what you think happened" and evidence as "what makes you think this is." The second participant expresses a view that we have seen repeatedly in our studies. Students using the SWH first make a claim and then are asked to find evidence. On some occasions, when searching for evidence, students realize they have made a claim that they cannot support with data. As the participant states, she has the "evidence for one thing and claiming for another thing," so students rethink and sometimes modify their claims to match their evidence. The third quote is an excerpt of writing from an SWH report written by an eighth grade participant. It indicates a strong connection in writing between evidence that is presented in the first sentence with a claim that is presented in the second sentence.

> [about the distinction between claims and evidence] Cause they're two different questions. And evidence is why you think or what makes you think this is . . . claim is what you think happened (boy, Ames study II).

> When you got to the evidence, you might think this isn't a claim at all and if you had to write down why it was then maybe your claim changed because you were reflecting on what you had done and then you realized that this is not the claim that we're saying. We've got the evidence for one thing and claiming for another thing, so you might change it (girl, Victoria study).

> *Well, our stream had 8.8 ppms of dissolved oxygen, or about 78% saturated. This is very good for the fish who live in the stream (boy, South Hall water quality study).*

Third, evidence from our studies indicates that using the SWH fosters the understanding that the claim must answer the scientific question. A girl from the Ames study, below, expresses the view that the question, "talks and it tells the claim," so she can readily respond to the question with an appropriate claim. Practice in coordination of claim, question, and evidence has also provided students with the notion that there can be more than one correct claim. For example, second participant quoted below talks about the possibility of a "perfectly legitimate claim" that doesn't answer the question and also about the experience of question formation, which was left up to the students in her classroom.

> I think that it [the claim] deals with the question and um that's like all because it just tells, it explains the questions more and like this question, it says what are the parts of the cell membrane? And it talks and it tells the claim- it tells what are the parts of the cell membrane (girl, Ames study I).

> The claim and evidence was kind of confusing to come up with. The claim too, because you might think the one claim, it might be a perfectly legitimate claim, but it might not answer the question you're asking. (Interviewer asks if she then changed her question). I didn't really change my question at all. I only had to change my claim a little bit . . . I tried to answer the questions I had some knowledge of so I wouldn't be completely in the dark to start with, but there would be something that I would be actually finding out that I didn't know (girl, Ames study I).

Limitations of the SWH: Working with Canonical Explanations

The SWH has promoted understandings of questions, claims and evidence for learners in our studies. When students generate claims and evidence for one laboratory investigation, it parallels the work of a scientist reporting on claims and evidence for a single experiment or field study. A difference between scientists' work and that of learners in the classroom is that the former approach most of their work with an explanatory frame already embedded into their knowledge structure. Students, on the other hand, are attempting to construct the explanations for their laboratory work at the same time as constructing an understanding of the nature of scientific investigation. This dilemma between investigating the authentically unknown and teaching canonical explanations may have fueled the confusion over the purposes and potential outcomes of laboratory instruction in general (Wellington, 1998).

> Assertion 4: The SWH has not consistently supported students in constructing or reconstructing explanations for their data.

Although the SWH student template encourages students to read relevant information in their textbook (including scientific principles, theories, or explanations) and compare their own laboratory findings with authoritative information, it has lacked scaffolds for student integration of authoritative knowledge with their own. We have found that students will sometimes create spontaneous explanations on their SWH reports in the evidence section, or the reading section. Often, students provide descriptions of their laboratory findings, rather than explanations. This is to be expected when the directive for claims and evidence, as it stated in the SWH, contains no prompt for explanation. Thus, the issue has arisen as to whether we want experiences with the SWH to include the consideration of explanations for scientific phenomenon as a way to provide pedagogical links between laboratory activities and science explanations. This issues includes a number of epistemological considerations, such as how students may view the link between a first hand observation and consensually constructed science explanations in a way that is more authentic than a simple verification.

One research study in particular, the Ames II study, illuminated this dilemma and provided data to indicate how individual student epistemologies interact with different ways of knowing in SWH experiences. This study investigated students' preferred knowledge sources (whole class discussion, textbook reading, peer group discussion, hands-on activity, SWH report writing, letter writing) in constructing understandings of osmosis/diffusion in seventh grade life science. In addition to preferred knowledge sources, the study investigated preferred cognitive strategies used by the students to interpret the knowledge sources and the nature of their osmosis explanations.

We found three distinct approaches to learning from SWH experiences. Three of six target participants utilized the first approach and were called *the first-hand observers.* These students relied on their hands-on activities and peer group discussion about those activities for constructing osmosis knowledge. Their preferred cognitive strategies were observation, listening to peer ideas, and explaining the information to themselves in writing. They did not use authoritative textbook information. Their written explanations for osmosis may be more accurately called descriptions. They reported on what they saw without trying to incorporate canonical science knowledge. For example, Christie's explanation below focuses on what she saw, including, how salt and sugar water made the cell walls darker, make them seem to move, and make the plant cell smaller.

> *"I think that the reason that the plant cells change is because of the different kinds of water that I used to find out the information that I needed. It also made some of the colors darker in some of the spots and it also made the cell wall somewhat smaller too. I think my answer is right because when we did the test it showed to me that salt and sugar can do a lot of stuff to plant cells like move the cell wall and make the plant smaller" (Christie, Ames study II),*

The second distinct approach was utilized by only one of the six target students, Allison. We called Allison *the reader*, because she relied on the textbook and information generated in whole class discussions to construct understandings of osmosis/diffusion. Allison purposefully did not use her laboratory observations as a source of explanation. She exhibited a sophisticated epistemological stance that while an investigator could see what was happening to the cells under the microscope, she couldn't link that observation directly with "why" it was happening. In other words, she asserted that you can't form an explanation from a singular observation. In her written explanation, she indicates how she relied on the textbook definition and the words of her teacher as her epistemological sources.

> *"My claim is that the processes that allow molecules to pass into and out of cells are osmosis, diffusion, active transport, and passive transport. According to the textbook and class discussions, directed and justified by Mr. M [the teacher], diffusion (the transport of molecules from high to low concentration), osmosis (the transport of water molecules from a high to low concentration), active transport (the transport of molecules using energy), and passive transport (the transport of molecules without using energy) are the processes which allow molecules to pass into and out of a cell" (Allison, Ames study II).*

The third distinct approach to interpreting knowledge sources was utilized by two participants, both high achieving students. Jason and Daniella were called *the integrators*, because they seamlessly blended first-hand observations and authoritative information from the textbook and class discussions into their osmosis understandings. These two students did not seem to be disturbed by the epistemological differences in observations and authoritative explanations. During her interview, Daniella asserted that she appreciated the chance to read osmosis explanation in her textbook and on the Internet, so that she could better understand her laboratory findings. The cognitive strategies used by the integrators included the widest range including observing, listening, questioning, hypothesizing, writing, reading and synthesis. In Daniella's written explanation below, the integration of what she saw in lab and the canonical explanation is combined to provide a rich and detailed account of osmosis.

> *"When a solution is added to Elodea plant cells the cells shrink and chloroplasts tend to move toward the center of the cell. The cell shrinks due to the method of substance movement called osmosis. Osmosis is the process of water molecule movement. When a cell is placed in an environment that has a low concentration of water molecules, the process of osmosis corrects that. Within the cell lies the vacuole and inside the vacuole is water. If the concentration of water molecules*

> *within the cell vacuole is higher than the cell's outer environment, the water molecules are released into the outer environment. The vacuole takes up the most amount of space within the cell. The vacuole pushes the cell wall helping it keep its shape, but when the water within the vacuole is released, the vacuole is deflated. The deflation of the vacuole causes the cell structure to shrink. Without the vacuole, the cell wall is not able to maintain large, rigid form it once had. Thus, the cell shrinks.*
>
> *When we added salt-water and sugar-water solutions to the cells they shrunk. This happened because the salt and sugar molecules take up a lot of space that is usually designated for the water molecules. With sugar and salt molecules hogging up the space, there are only a few water molecules present. The cell realizes this and since the water concentration within the cell is greater than the outside environment, the cell goes through osmosis. This could also explain the rapid movement of the chloroplasts when we added the sugar-water solution" (Daniella, Ames study II).*

CONCLUSIONS AND IMPLICATIONS FOR FUTURE RESEARCH

Research to date on the use of the SWH in several classrooms has indicated positive results. Quantitative results have shown that, overall, students using the SWH score significantly better on higher order conceptual questions of science knowledge. Qualitative results have indicated that the SWH promotes metacognitive thought about a laboratory activity, reflection on scientific concepts, and improved understandings of the nature of science, including questions, claims, and evidence. Thus, we assert that the use of the SWH in science classrooms will promote important components of scientific literacy.

It is important to note the limitations of our studies, and areas about which little is known. For example, not all of the quantitative studies of the SWH resulted in significantly better scores for SWH treatment students. In the Ballard study, potential factors such as the nature of the students, the tasks, or teacher instruction may have rendered SWH treatment less effective. In terms of the qualitative findings, we do not have evidence that all students experience metacognitive thought during SWH activities. In the South Hall erosion study for example, the SWH prompts promoted the search for hypothesis and evidence among some writers during writing, but not all. We still do not know how other students are making scientific and cognitive connections during SWH activities or whether they might make the same connections with traditional instruction.

Similarly, we have little information on specifically how students might change their science conceptions through SWH experiences. Is the consideration of new evidence in their laboratory activities an important component? Do the activities of writing or reading promote or solidify new views? How widespread is the change in science conceptions through the use of the SWH? The whole area of conceptual change in relation to SWH activities needs additional research. Also, as indicated in the final results section of this chapter, we need more research on how to integrate canonical science explanations with first-hand observations in SWH classrooms. We are currently developing a new modification of the SWH that prompts students to create or find an explanation for their laboratory results.

Finally, a substantial limitation of our studies has been the narrow demographics of our research sample. The participants in our studies have been largely White and middle class. Research on the SWH needs to be expanded to include students of other ethnicities, socio-economic groups, and learning styles. The development of SWH activities for students in urban settings and those with large English language learners populations is greatly needed. Research on these more diverse settings is necessary to determine how all science learners are responding to the SWH.

APPENDICES

Appendix A: Interview Questions for the Ames, Ballard, and Victoria Studies

1. You have shown improvement from the pre test to the posttest. What do you think helped you learn this topic?
2. Which activities during science lessons do you find most helpful, hands-on activities, reading, class discussion, small group discussion, or writing? Why?
3. What kinds of things do you talk about in small group discussion?
4. Do you believe writing helps you to learn? Why?
5. How does this science class compare to other science classes you have had in the past?
6. How did you come up with your claim (question and evidence)? Was it hard or easy, why?

Appendix B: Interview Questions for the South Hall Water Quality Study

1. Considering you were scientists on the stream study, how do scientists work?
2. Do scientists always agree, why or why not?
3. What do scientists do when they disagree?
4. What did your group do if you disagreed?
5. How do scientists' ideas change?
6. What do scientists do with their data and results of their tests?

7. Why do you think scientists write reports?
8. What new things did you learn about how scientists do their work from the stream study?

Appendix C: Interview Script for Think-Aloud Protocols

1. Introduce the session. "Hello, (student's name). Thank you for agreeing to participate in this research study. Today's session will last about two and one-half hours and you will have at least one break. The purpose of this research is to find out more about what is going on in students' minds when they are doing science writing. In order to do that, we are going to ask you to think aloud during your science writing today. Think aloud means that you say out loud whatever comes into your mind as you write. We will practice on a familiar topic first. After we practice thinking aloud, we will do a science lab on erosion together and gather some data. When we finish the lab, you will talk together as a group about your lab result: what you found out in your experiment. Then, after a break, you will write a science report about your experiment, thinking aloud while you write it. Do you have any questions right now? You may ask questions at any time."
2. Practice thinking aloud. "Now we will practice thinking aloud while your write about your favorite pet, person, or hobby. I would like you to write a brief description of your favorite pet, person, or hobby and think aloud by telling me everything that comes into your mind while you are writing. Please say anything that comes into your mind, even, "I don't know what to say next." I will remind you to keep talking from time to time, but I will not tell you anything to write. This practice session will last about 10 minutes and be recorded. (Escort students to separate areas of the room where they will not disturb each other and turn on tape recorders.) You may begin."
3. Doing the erosion laboratory activity. (Bring students back together as a group of three. Review the meaning of the term erosion and describe the erosion lab, as it would be done in a science class. Give directions and explain use of materials. Escort students outside to carry out the erosion lab.)
4. Discussing erosion laboratory results. (Gather students in a group with their data sheets. Allow them to discuss their results with the use of focus questions on the lab handout. Videotape discussion.)
5. Completing the think aloud session for report writing. "Now that you have done some research on erosion, I would like you to write a one to two page report on your findings. Use your data to remind

you of important points. Do the best you can. While you are writing this report, say out loud everything that comes into your mind. I will remind you to keep talking. You may take as long as you need, but you will probably be done in 40-50 minutes. This think aloud session will be tape recorded." (Separate students once again. Turn on recorders. Provide writing prompt guideline, paper and pens.)

Appendix D: Writing Prompts for the South Hall Erosion Study

Thank you again for volunteering to do this project! The last step is to write a report about your erosion investigation and think aloud as you are writing it. As you write your report, make sure to include the following:

1. Your scientific opinion of how bad erosion is behind the tennis courts a South Hall Middle School.
2. Detailed evidence supporting your opinion, including the results of specific observations and measurements you made.
3. A description of how you carried out the observations and measurements.
4. How you findings compared with your predictions about erosion at South Hall.
5. Possible causes of erosion at South Hall.

CHAPTER 7

CHILDREN'S VIEWS OF WRITING TO LEARN

Carolyn S. Wallace & Brian Hand

From the earliest beginnings of the writing to learn movement, the case has been made that because writing is a verbal, visual, and kinesthetic activity, it involves the learner in self-discovery of knowledge. As Britton (1970) originally suggested, writing allows us to "explain the matter to oneself." This unique learning feature has the potential to be very powerful in generating new understandings. Many studies of writing to learn have included testimonials of students, expressing their positive views of the process (Conally & Villardi, 1989).

Almost universally, qualitative, anecdotal, and statistically based reports of what students think about the value of writing for learning have been positive (Rivard, 1994). Mason and Buscolo (2000), for example, investigated change in 11 and 12-year-olds perceptions of writing, as well as their science conceptual change. They found significant differences in post writing instruction measures of perceptions such as using writing to "reason and reflect on new things", and to "express what you are not able to understand" for children in a writing experimental group, but not a control group. Similarly, children in the writing group indicated that they found all four types of classroom writing in the study to be useful, whereas control group children did not.

Our own research studies support this general trend and have aimed to delve somewhat more deeply into specifically why science learners find writing helpful. During several of our studies we have interviewed students concerning their perceived values of the writing tasks. Our research questions have included: (1) Which types of writing tasks do students find most useful in science learning? (2) How does transforming writing for an audience other than the teacher influence students' writing from their own perceptions? (3) What types of cognitive and metacognitive processes do students identify as being used during science writing? In this chapter, then, we present a synthesis of these interview studies, including data from 55 students.

The theoretical framework for the synthesis of studies in this chapter derives mainly from the seminal work of Emig (1977), described in Chapter 3. To briefly review, when a writer makes thinking visible, she can identify gaps in her knowledge base. This identification of gaps may cause her to go and seek out additional information and make new connections. She may also ask new questions about the topic. Writing will require the learner to organize the piece into a linear progression, which involves the generation of a logical sequence. When she explains an idea in a piece of writing, she is required to make connections between discrete pieces of information. And when she looks back over what she has already written, she may generate even more connections or be pressed to search for new information (Emig, 1977; Langer & Applebee, 1987).

METHODS

The research design for this study is interpretive in the broad sense, deriving from Erickson's (1986) view of interpretive research as gaining meaning from the participants' point of view. We interviewed children to determine how they viewed the various writing tasks posed to them in the classroom, whether they thought the tasks had value and why, and how they viewed the writing tasks as fitting in with the overall teaching and learning agenda in the classroom. All of the interviews were semi-structured with several pre-determined questions and many probing questions. Representative interview questions are shown in Appendix A.

Five representative studies of children's views of writing are included in this chapter. These include the Ames Middle School studies (both I and II, Hand, Wallace, & Yang, in press, Wallace, in press), the Ballard study (Tsoi & Keys, 2001), the Victoria study (Hand, Prain, & Wallace, 2002), and the Central Iowa study (Yang & Hand, 2000). The context of the first four studies and the selection of interview participants have already been described in Chapter 6. To summarize briefly here, students in the first four studies were studying seventh grade life science, tenth grade biology, and ninth grade physical science, respectively. The two Ames studies and the Ballard study took place in middle socio-economic suburban schools, and the Victoria study in a rural Australian secondary school. The study participants were overwhelmingly White. Students in these four studies had participated in at least two of the following kinds of writing activities, SWH laboratory reports, letters written to peers or parents, and textbook explanations written for peers or younger students.

All of the students in the Central Iowa study were taught by one teacher. The teacher was a White male with seven years teaching experience. The subject matter for the Central Iowa writing study was stoichiometry. This topic was explored both computationally and conceptually through the use of science writing. Students used writing to learn strategies, including explaining the process to a younger student, instead of end-of-chapter questions from the text. Nine students were interviewed after the completion of the final test. These students were selected to include balanced gender representation and a range of science achievement as determined by course grades and teacher judgment. The interview questions focused around the two major ideas: (a) What did students need to do to successfully complete the task? And, (b) What value did the students gain in completing the task?

All interviews were audiotape recorded and transcribed. Data analysis consisted of unitizing the transcripts into single idea units, coding the units with descriptive and interpretive codes (Miles & Huberman, 1994), collecting the codes into categories, noting the most frequent codes and categories, and developing assertions from ideas that emerged repeatedly (see also methods in Chapter 6).

RESULTS

Types of Tasks Students Find Useful

A frequent question in studies of writing to learn science is what types of tasks evoke the most thinking and reasoning on the part of the students. Langer and Applebee's (1987, see Chapter 3) pioneering study found that writing tasks that required synthesis, such as writing an essay, were more effective for deeper thinking and longer term retention than more rote tasks, such as notetaking or answering textbook questions. While our investigation of writing tasks is constrained by the activities we asked students to engage in, including SWH reports, textbook explanations, and letters, students were able to identify three types of tasks that helped them learn most.

> Assertion 1: The types of tasks students found most useful in helping them learn were framing their own questions, explaining a process, and rewording science information for a particular audience.

The activity of framing science questions in writing took place in classrooms where we implemented the SWH. The seventh grade life science teacher in the Ames studies, in particular, was very skilled at working with his students to frame their own questions for laboratory investigations. He used a variety of techniques including group brainstorming, and whole class discussion to stimulate the generation of questions. In the Ames studies, several students expressed question framing as being a valuable learning activity. In the following interview quotes, the children expressed why they believed framing a question was valuable. Their comments point to the ideas of increased interest (first and second quotes), providing focus for learning (third quote), and greater exploration of science topics (fourth quote).

> It makes you more interested 'cause like if you have to answer something you're not interested in, you already know it or something, you're bored and you don't pay attention, but if you answer a question that you made up, helped make up, and you're wondering about, you pay attention, cause you want to find out the answer to your question (girl, Ames study I).

> [Which one do you like better, traditional labs with one question or SWH where you choose your question?] I guess the SWH but where you get the different questions because you can pick your own. Yeah, because if it's something, I guess if there' something I'm interested in, I can pick a question about that and then I'll learn more about that (girl, Ames study I).

> Well, I thought it was kind of mostly all of it [portions of the SWH] because you stated like a question about something and then you learned specifically about what that does and it helped you a lot to learn (boy, Ames study I).

> I like this class more than any other science classes I've been in or any studies, because I just think the learning is a lot more extensive and we don't just answer one set question. We'll answer that and then we'll explore all the sub topics and everything that has to do with that question. It's like the mind maps because they branch out (girl, Ames study I)

The second type of writing task students identified as most useful was explaining a science process. Explaining a science process took place when students wrote textbook explanations for peers or younger students, letters to parents, or in the SWH reports. Explaining a science process in writing helped students mentally engage the steps of a process, like solving a stoichiometry problem. It also helped them identify their own understandings, and express their understandings in words. In the following interview excerpts, students stated their ideas about why explaining a process is useful, including reviewing the steps of a process (first quote) and synthesizing information (second and third quotes). In the fourth excerpt, the student is ambivalent about the value of explanations, because writing explanations was difficult for her and revealed gaps in her knowledge.

> [The process] was definitely good for me to do. Just to help me better understand what I am talking about because, I was going through the steps of stoichiometry in my head (girl, Central Iowa study).

> [What helped your learning?] The writing on the cell. I just thought it was easier for me to understand when I wrote it down in my own words. I have to look up stuff and dig deeper into things and figure out how this cell works (boy, Ballard study).

> I had to take all the information I got and I had to put it all together and explain . . . I had to take my information and explain why I came to the claim I did (boy, Ballard study).

> It wasn't really useful for me because I couldn't explain my ideas very well. I'm not very good at explaining things to make them understand. There is like a lot of organo's and I couldn't write all of them down . . . I can't explain them all. These are just the ones I learned about. It was helpful. Not a lot (girl, Ballard study).

The third type of task that students found most helpful for science learning was rewording their science ideas for a particular audience. During our studies, we varied the audience for writing to determine how expressing science ideas in different language modes affected students' cognition. The most frequently used techniques were writing for parents, writing for peers, or writing for younger students. We hypothesized that writing in different modes would make students develop flexibility and applicability with their science understanding. Findings for the effectiveness of writing to a different audience led to the following main assertion.

> Assertion 2: Students believe that rewording scientific content into forms suitable for parents or younger students requires language transformation that increases understanding.

Students from grade 7-11 easily recognized the need to change the language for an audience other than the teacher. Students repeatedly articulated the understanding that the teacher already has an in-depth knowledge about the content, and so they could use terms and explanations in more of a rote manner, without their own complete understanding, and still be confident that it would make sense to the teacher. When writing for a different audience, the scientific language needed to be unpacked and explained using more everyday language. This activity of unpacking the language triggered metacognitive thinking on the part of the students, as they were required to verbalize their thoughts in more than one language representation. The quotes from children below illustrate their understandings of: (a) the teacher's knowledge (first, second and third quotes); and (b) the need to use transformed language for an audience unfamiliar with the scientific domain (first, second, and third quotes).

> . . . we wrote to somebody else instead of Mr. M. and so you had to put your ideas in a different way so that someone your age who hasn't

> studied it could understand it . . . I think I had to put a little bit more effort in it because Mr. M. was studying the thing with us and he was teaching us so he knew what we were trying to say and when you're writing to someone who doesn't understand anything you have to go into really specific details for them to understand (girl, Ames study I) .

> It is harder to write for another person than a teacher because it's another person your age and you have to speak their way, whereas the teacher will accept any sort of words (boy, grade 9, Victoria study).

> The vocabulary [for a younger audience] has to be a lot easier to understand. When you are writing to a teacher, they already know what the stuff means. You are not trying to tell them what they [the words] mean; you are just trying to use them to tell that teacher that you know what they mean (boy, Central Iowa study).

Feedback from the peer or younger students, and then redrafting the textbook section or the letter based on the feedback was also very helpful to students. Especially in the Central Iowa study, when students were trying to explain stoichiometry to seventh graders, initially they had little idea of the seventh graders' prior knowledge. When they received feedback, the chemistry students found specific areas of weakness in their explanations. Writing for an audience requires a clarity of understanding as expressed by the following student.

> The math changed a little bit because I messed up on the math and I didn't even realize it until after they wrote back and said that they didn't understand the math. I read it and it wasn't right. I did the math wrong so I had to change that and I explained what ratios were a little bit clearer (boy, Central Iowa study).

Scaffolding the Writing Process

In the studies that provided planning support to students before they completed the writing tasks, student interview data revealed some important patterns about planning. These studies were conducted at Ballard and involved treatment students receiving scaffolded support prior to writing, while control students received planning support prior to completing a re-draft of their text.

> Assertion 3: The planning support provided was beneficial in helping the students understand, and be prepared, to complete the writing task.

The full writing support provided before the first draft for the Planned (PL) group and before the second draft for the Delayed Planned writing group (DPL) group was viewed by the students as beneficial for completing the task. When asked if they felt prepared when completing their first draft, 58% of the PL group said that they were prepared compared to 25% for the DPL group. The difference can be noted by comments such as those by Chris in the planned group who said "We had all the information. I was totally prepared" as opposed to those by Amy (DPL) who said, "I think the first draft was really hard.... I didn't know what I was doing".

The planning process was viewed as valuable because it helped students organize the task and determine where they needed information. These ideas were expressed by Jim (PL),"it helped to organize it [the writing], and Cara (PL), "once I used the wheel [template] before I researched I think it helped because that way I knew exactly what portions of the research I really needed." This confidence in understanding the task was not expressed by those in the delayed planned group; for example, Shelley (DPL) stated, "I didn't really understand exactly what I had to write. I didn't understand that I had to put in so much detail", while Chantel (DPL) said "you don't know what order to put it in and stuff like that...[I was] not very prepared at all. I mean I didn't know anything about it". The planning sessions for the DPL group enabled them to go "back over it [the first draft] and kind [of have an] overview. We looked at all those questions and actually saw what to look at and improve," (Heath, DPL).

While nearly all the students liked the planning opportunities when presented to them, Jamie (DPL) was not one of them. For him, writing does not require such planning "because I don't outline the writing. I just kind of write, then I go over it. I really didn't find it [the planning sessions] all that useful". Both control and treatment groups had the opportunity to present and receive feedback on their ideas from the whole class, although the planned group (PL) got feedback on just their initial ideas, while the delayed planned group (DPL) got feedback on their first drafts. The opportunity to receive feedback from the class as whole was seen as being valuable by both groups and 92% of the students in each class stated that the timing was appropriate. However, students in the DPL group indicated that as they presented their first drafts for their class presentation, "everyone could help each other out to make it better, since we already wrote it. It helped us to make it better, make sure it was understandable," (Shelley, DPL). The support from each other as a consequence of the presentation for the DPL students is reflected in Keith's (DPL) comments when he stated:

> Yes, because I had the chance to see what other groups were doing and how they explained stuff. They kind of put it into simpler terms than I did and that made me change mine so I could explain it better and so I could understand it better. They had little bits and pieces just like I did and I put the whole big picture together with the group.

For the PL group, the presentations were able to help students determine the amount of information that they had collected or needed to collect. For example, Chris (PL) said that "it helped us think more and get more information so that we could make our papers better, so we could put more information on there". This feedback gave students "more options with what to do and what you needed to work on," (Brooke, PL) or that other students had "found something that you didn't" (Jeff, PL).

> Assertion 4: Social negotiation of meaning was perceived by the students to be a critical component of the planning process.

All (100%) of the students interviewed in both groups indicated that working within their small groups was helpful in constructing understanding. The focus of these negotiations was predominately around either the information required in the writing or the appropriateness of language required for the younger audience. Using negotiation to construct meaning was perceived to be useful not only "because you all share your ideas and it makes it easier" (Chantel, DPL), but because there were also opportunities where " if you didn't catch something someone else caught it" (Katie, PL). As Jamie (DPL) pointed out "if you have somebody else who is doing the same thing, then you can kind of get some of their information and compare it. See how you did" and this helped generate confidence in their understandings because "you just knew" (Katie, PL) what was supposed to be put into the paper. The knowledge constructed was viewed by the students as having to be more than just their own thoughts but rather a mixture of their own and the group responses to their ideas. As Cara (PL) said,

> When we all got together and discussed this is my research, how does it relate to yours, then when people would bring up these different topics there is kind of that point where somebody would say no that's not right or that's not what I think but they figured out a way to word it so it combined the two into the paper.

This opportunity to add to one's own knowledge was viewed by Jimmy (PL) as really important for his learning. As he said,

> Discussing it helps me a lot that is one of the main things that helps me. By writing it down and reading it, it helps but not as much as discussing it because you can get everyone's opinions and put your own into and see if you are right about it.

The value of the negotiation process was not just having the information correct but the opportunity to have feedback on the level of the language used for the Year 7 audience. By discussing their ideas within the group, students were able to understand that "we just needed to put it in easier words" because "if they [the group] didn't understand it, I knew I had to go back to my paper and change it," (Shelley, DPL). In responding to the group's suggestions, students were able to complete what they perceived to be a better product. As Jim (PL) pointed out:

> Some people said to change the wording, so I did that. I used smaller words so that they could understand what I was talking about. And I lengthened paragraphs since I had added stuff, and took some stuff out. [I] sorted it out and made it better

Social negotiation was a critical component of the planning process. Having "the 4 brains or 5 brains or however many," (Matt, DPL) working together was beneficial in helping students plan and write their textbook and newspaper articles.

> Assertion 5: Students in the PL group were engaged in a deeper form of forward searching then DPL students when redrafting their pieces of writing.

Analysis of the interviews did indicate that there were differences in the actions taken during redrafting between the DPL and the PL group. Students in the DPL group predominately referred to the need to add more words of explanation, that is, they did not feel as though their first draft was adequate in terms of the explanation provided. For Amy (DPL) there was a need to add more because,

> The first one I didn't know what I was doing. I was just writing stuff down. On the second one I think it helped me understand a lot more because I realized that I didn't put enough information into the first one and I wouldn't even have understood it if I was a 7th grader reading it. So I went back and did a better job on it.

> The concept of having to add more information was centered around what Matt (DPL) referred to as the "surface" of the subject. He said 'I didn't get too in-depth in the first one. I kind of just hit the surface slightly. On the second draft I kind of got into it". Keith further expands on this point when he said, "I just needed to elaborate more and continue on with my ideas. I wasn't expressing enough". The greatest change in completing the task was by Skylar (DPL) who "changed my whole paper".

Students in the PL group consistently referred to the opportunity to re-order their ideas when completing their second draft. Not only did they state that they re-worded their ideas to deal with the audience, but they were able to make better links between the ideas and thus construct a better sense of organized conceptual framework for the topic. For Courtney (PL) it was a matter of getting "the sections and stuff organized better so it is easier to understand. There is one section about something and the next section will be about something, so that you don't mix it all together. It is all like separately organized." The redraft process enabled these students to organize their understandings into a more coherent and connected set of ideas. As Brooke (PL) said:

> You have to be organized. You have...I mean you can't write about something else and then explain it like further down on the page or tell about something and not tell what it means. You have to make sure you are going in the right direction with it so they understand what you are talking about through the whole paper.

The DPL group did not refer to the organization of their concepts in the manner in which the PL group did. These students tended to be at the initial stage of cognitive organization; they were still dealing with goals of the tasks as well as the conceptual explanations. The PL group were focused on the relationship between the concepts and how best to represent these and thus were engaged in a deeper level of forward planning. This deeper form of forward searching is reflected in the results from the quantitative component of the study.

Thinking Processes Identified by the Students

Students in our studies identified both cognitive and metacognitive values of writing to learn science. Some of these, such as increased understanding through language transformation have already been characterized above. Two additional ideas expressed by the students for the value of writing are discussed in assertions six and seven below.

> Assertion 6: Students characterized writing to learn as breaking through the superficial cognitive activity of memorization to deeper modes of thinking and making meaningful connections.

Often in our interviews, students mentioned how they could not memorize the content material and do well on the writing to learn assignments. These students seemed to come to a new understanding of what it means to learn something in science. As stated by one participant, "I didn't always get the best scores, but it helped me learn a lot more" (boy, Ames study I). In the interview excerpts below, students point out how writing helped them (a) avoid memorization (first quote); (b) go beyond plugging numbers into an algorithm (second quote); (c) think more deeply (third and fourth quotes); and (d) generate authentic learning (fifth quote).

> Well then, I think this helps you out more because memorizing sometimes you don't really learn what it actually means and then but learning it and remembering it instead of memorizing, it makes you learn more, I think (girl, Ames study I).

> When we started this, I understood how to do it, but I didn't know why to do it or it was just kind of like an equation and you plug numbers in. When you explained it, you kind of had to tell why you were doing everything so that it would make sense to them (girl, Central Iowa study).

> I think it really makes, specifically it really makes you think. And that's probably the thing that he's, Mr. M has taught us the most. You have to think, and you have to push yourself to ask more questions and to get more information and to just really look deeper . . . And so that you can explain what happened or why it happened (girl, Ames study II).

> The [SWH] template made you think about what you saw, what you saw does, and other things I had never even thought about (girl, Ballard study). Like, it doesn't really show you've learned a lot by answering the textbook questions because they're in bold. You don't even have to read it you just copy down what's in bold and I know a lot of people do that. But you can't learn if you do that and then when he [teacher] assigns things like the paper and the SWH, you can't fake it, you have to really know what you are talking about (girl, Ames study I).

> Assertion 7: Students identified writing to learn as a metacognitive tool for self-regulation of learning.

Finally, students in our studies identified the value of writing for recognizing gaps in their own understanding. They viewed writing as a check on their own learning. Several students talked about how they didn't really know their own level of understanding until they attempted to explain it in writing. Thus, our studies support the long held hypothesis that a major value of writing to learn is metacognition. In the following quotes, students illustrate two main points. The first point, illustrated in the three quotes shown below is that writing helps you see what it is that you don't understand. Finding gaps in knowledge prompts one to regulate their learning, as the participant quoted first indicates, "it will tell you [that] you need to learn more."

> Yeah, because if you don't really know something very well then if you try to write about it, it will definitely show. I mean it will tell you [that] you need to learn more so then you go back and check (girl, Ames study I).

> . . . I explained it the way if someone said this, like I would write down my thinking and go and read and make sure it made sense to me and then I had my mom and dad read it and make sure it made sense to them too
> . . . [writing helped content understanding] because otherwise [if I don't write it] I think I understand it, but then when I write it down, I understand that I don't understand the topic (girl, Ames study I).

> I started writing the letter and I found out that I couldn't really write about it because I didn't know about it. I don't think I understood it well enough to explain it (boy, Ballard study).

The second point under the assertion for writing as form of self regulation is that students believed the writing activities in our SWH studies promoted autonomous learning. The SWH activities were based more on open-ended inquiry, so that students were often required to frame their own questions, then design ways to answer their questions. Accountability for their learning, as they composed SWH reports and final projects may also have promoted independent learning. As one of the students below stated, "we found out how to learn."

> We do a lot of stuff ourselves, the teacher is just there to guide us (boy, Ames study I).

> Well, I think probably the labs . . . you actually had to find out for yourself so that way we, not only did we learn, but we found out how to learn (boy, Ames study I).

> It [answering SWH questions] was okay, because you just have to know what you need to learn and if you don't know it, you learn it. I thought it was good to answer (girl, Ballard study).

CONCLUSIONS AND IMPLICATIONS

This synthesis of children's views of writing provides additional evidence for the prevalent hypothesis that learners recognize the value of writing for learning. Our studies indicate that children as young as seventh grade think insightfully about the meaning of their own schoolwork. More specifically, this synthesis has contributed to our understanding of why students value writing and what types of task are viewed as most useful.

To summarize, the types of tasks students find most useful include framing their own questions, explaining a process, and reframing content for an audience other than the teacher. Students enjoy posing their own questions for investigation in science class because they have the opportunity to learn material that interests them, and they have a clear focus to direct their learning. When students are able to form their own questions for investigation, they also develop a sense of autonomy; this is their learning, not the teachers' learning. As one participant expressed, the unfolding of questions leads to exploring extensive networks of learning, which she enjoyed. The implication for the classroom is to provide more opportunities for students to pose their own questions for learning, either for laboratory investigations or investigations of print-based material. When teachers become skilled at coaching the posing of questions, students can develop a sense of ownership over their learning, even when the topics are mandatory for the curriculum.

Second, students found explaining a process to be a particularly useful writing task. We may speculate that during this kind of task students are required to organize their thoughts in a linear manner, think logically through the process, discover gaps in their own knowledge, and make meaningful connections among disparate pieces of information. These results support many previous theoretical statements (Britton, 1970; Emig, 1977; Langer & Applebee, 1987) about the power of writing to learn. Explanation is indeed an important genre of science writing (Keys, 1999), which is probably underutilized in science classes. Students may be engaged in: (a) explaining how to carry out a problem-solving process (i.e. stoichiometry); (b) reconstructing for themselves theory that they have read about in the textbook (i.e. explaining the process of blood circulation; Santa & Havens, 1991); or (c) patterns in their own first hand observations (i.e. how the shape and size of gullies lead to inferences about erosion (Keys, 2000).

Third, students find reframing content for another audience to be useful in constructing science understandings. When involved in this task, students recognize they cannot use the same technical language they would use for the teacher as audience. Therefore, they are required to unpack technical language into everyday language; they must make explicit and detailed connections; and they must understand the theoretical basis for the words they use. The implication for the classroom is that students need opportunities to write for multiple audiences. Explicit instruction on writing for a different audience may help students understand the goals of this type of writing. The positive comments made by students on writing for a different audience in our study reinforce the idea that learning science requires a translation between everyday discourse and scientific discourse.

Students also found that planning tasks were helpful for writing. They expressed that presentation of their ideas to the class, feedback from the audience, and peer group negotiation were especially helpful. Those who participated in advanced planning, that is before writing an initial draft commented on a deeper level of forward search processes than those who planned only for a final draft.

We also understand more about why students value writing as a result of this research. Students appreciate the distinction between rote learning and meaningful learning. They like having the opportunity to make connections, to think deeply about what they have observed, and to attach their learning to their own questions and prior knowledge. Students are also positive about the metacognitive benefits of writing to learn. Students seem to value the opportunity to think on their own and express science in their own voice. In short, they enjoy the interactive-constructivist patterns of learning, just as science educators do. The implication is that more meaningful writing activities should be incorporated into science classrooms. Teacher education courses should have substantial time devoted to the pedagogical practices of writing to learn.

The most important limitation of our studies thus far is the lack of diversity among the students with whom we have worked. Although representing both suburban and rural populations, our participants have largely been White and middle class. The perceptions of writing to learn of children from poor backgrounds, urban centers, multiple ethnic groups, differentiated genders, and those with learning disabilities is sorely needed. Future contributions in this area will provide necessary data to continue to promote writing to learn strategies in science.

APPENDIX: SAMPLE INTERVIEW QUESTIONS

1. Which writing activities helped you learn the most?
2. Did writing the SWH reports help you learn? How?
3. Did writing the claim and evidence sections help you learn? How?
4. What was it like to write a textbook explanation for seventh graders?
5. What differences were there in writing for a seventh grader and writing for your teacher?
6. What kinds of things do you talk about in your SWH groups?
6. Does talking to your group help you learn? How?

CHAPTER 8

TEACHERS' PERCEPTIONS OF WRITING TO LEARN STRATEGIES

Brian Hand, Vaughan Prain & Carolyn S. Wallace

Chapters 5, 6 and 7 have outlined the implementation of writing to learn strategies within science classrooms and some of the learning benefits gained by students. This chapter focuses on the role of teachers, who are an integral component of any classroom environment and have an obvious impact on the success of such strategies in terms of student learning. We have previously discussed that writing is a process of negotiating meaning in which there is both a personal and social component. As such there is a vital need for teachers to adopt pedagogical practices that enable students the opportunities to engage with these negotiations and to meaningfully challenge their own conceptual understandings of the topic. The implementation of writing to learn strategies as described within this book requires teachers to understand that learning is not about demonstration of knowledge, but the active construction of a deep and rich understanding of the conceptual frameworks of a topic.

This chapter describes two studies that were undertaken to determine the issues and concerns that teachers had in implementing writing to learn strategies within their classrooms. In each of these studies there was recognition by the teachers of a need to adopt new pedagogical practices if they were to successfully implement writing to learn strategies within their classrooms. Each of the studies was based on a long-term professional development program that was centered on examining the impact of using writing to learn strategies within school science classrooms. Both of these studies were based on qualitative research methods. The first study was conducted during the early research that we conducted on writing to learn strategies within classrooms and was focused on the struggles that two experienced teachers had in implementing these strategies for the first time. The second study was based on the implementation of the Science Writing Heuristic (SWH) as a means to use writing to assist student learning within inquiry-based laboratory activities. The three teachers involved were implementing the SWH approach for the first time. Both of these studies involved a partnership between the schools and the university involved. A long term approach to professional development needs to be recognized as an essential element in the successes achieved in implementing the writing to learn strategies.

THEORETICAL FRAMEWORK

Currently there is broad agreement about conditions that enable and sustain effective teacher change. Research on successful, enduring teacher innovation (Bell & Gilbert, 1996; Caldwell & Spinks, 1991; Crump, 1993; Fullan, 1994; Hall & Hord, 2001; Loucks-Horsley, Hewson, Love & Stiles, 1998; Tytler & Waldrip, 2002) has identified the following specific conditions as likely to promote this change: a clear well-defined set of goals for the innovation; a culture of teaching staff collegiality, and commitment to and ownership of the innovation; and effective sustained teaching staff professional development to support engagement with, and commitment to, changed practices. As noted by Tytler and Waldrip (2002, p. 23), long-term professional teacher development can only be achieved by programs that are "sensitive to the needs of teachers and schools." Hall and Hord (2001), and others, have emphasized the key importance of extended opportunities for teachers to discuss and influence the nature of the ongoing developmental program as a key factor in successful teacher change. By contrast, there is considerable research evidence of the ineffectiveness of short, transmissive inservice programs by 'external experts' that offer context-free renovations of teaching practice (Levins, Pegg & Creedy, 1994; Stronck & Koller 1991). Sound professional development needs to be built on sustained opportunities for teachers to increase both their disciplinary knowledge as well as their pedagogical content knowledge as they seek to integrate curriculum content and meaningful learning experiences for their students (Novak, 1998).

In terms of changing teachers' practices in science teaching, Loucks-Horsley, Hewson, Love, and Stiles (1998, p. 1) asserted that teachers need "new knowledge, skills, behaviours, and dispositions." They asserted that teachers need to focus more on methods such as, "inquiry-based learning, investigations, problem-solving, and applications of knowledge" (p. 36), on the assumption that past transmissive modes of instruction have proven less successful in engaging learners. These researchers also claimed that teachers need to feel competent and comfortable in their capacity to create new appropriate learning environments for their students. They also noted that teachers' knowledge is mainly "embedded in their practice rather than in codified bodies of knowledge, and that their extensive, complex knowledge, particularly with respect to their understanding of how learners learn, profoundly influences how they teach" (p. 32).

In other words, teacher change is more likely to be prompted by positive teaching experiences with new practices rather than by exposure mainly to new theoretical considerations. Loucks-Horsley and her colleagues (1998, p. 39) further claimed that "teachers were likely to change their attitudes if they saw students benefiting from a new practice, where instead of being linear, changes in ideas and attitudes, and actions and behaviors, occur in a mutually interactive process." They asserted that well-conceptualized professional learning experiences necessarily promoted changes in both beliefs and actions, where "change in one brings about and then reinforces change in the other" (p. 39). These writers also claimed that teachers needed extended engagement with new practices over time in a nurturing, collaborative environment if real and durable alterations were to occur in both teachers' beliefs and practices. In this view, teachers need sufficient opportunities for thorough investigation of the effects of the proposed changes on students' learning as well as their new classroom role, so that they can connect and, where appropriate, integrate new approaches with their current professional experiences and practices.

There is also an extensive research literature on the sequence of concerns teachers are likely to experience as they participate in change processes associated with professional development (Hall & Loucks, 1979; Hand, 1996; Loucks, 1996; MacDonald, Atkin, Jenkins, & Kemmis, 1997). From this perspective, teachers' initial concerns usually focus on the perceived personal impact of the innovation on their own self-perception as teachers. The next set of concerns will focus on task-oriented issues relating to efficient implementation of the innovation, including practical and organizational matters. In this stage teachers are often concerned with following perceived explicit or implicit guidelines or goals entailed in the innovation. Finally, teachers are then likely to be concerned with evaluative issues such as gauging the effectiveness of the innovation in terms of student learning and attitudes to the subject or topic. After these earlier concerns are addressed, teachers are then likely to be in a position to consider the possibilities of customizing the innovation to suit their own beliefs, style of teaching, and perception of students' needs. Loucks (1996) noted that this sequence of concerns has significant implications for professional development in that each issue or concern needs to be understood and addressed when it arises, or at the appropriate stage, and that sufficient time, measured realistically in terms of years rather than weeks, needs to be allowed for all concerns to be addressed effectively. This concerns model also acknowledges that individuals will have different needs at various stages of engagement with the innovation, and that effective professional development work must address these individual differences in teachers' needs effectively. In summary, the literature indicates a range of specific conditions need to be met for a professional development program to lead to effective teacher change, and also indicates the kind and sequence of responses teachers might be expected to make in engaging with any proposed innovation.

STUDY ONE

Methods and Context

This study involved two teachers: Alan, who had 20 years of experience, and Chris, who had 10 years of experience, both White males. Both teachers had participated in a long-term (four year) inservice program looking at the development and implementation of writing to learn strategies within the science classroom conducted in Bendigo, Australia. A total of 10 teachers were involved in this program over the four year period. In each year of participation, Alan and Chris implemented a number of different writing assignments within their classes including developing brochures, travelogues, letters to students in other settings, and letters to local farmers and council members. Importance was placed on using a difference audience than the teacher to write to, and to involve the audience in some form of review of the products. The tasks were focused on asking the students to explain the big ideas framing the unit under study. During the period of the inservice program, the teachers were supported through discussion of classroom observations, modeling, regular meetings of all teachers, and planning support where needed. Both teachers had previously been involved in an 18-month program aimed at implementing constructivist teaching/learning strategies within their classrooms. Data sources for this study included classroom observations, field notes, semi-structured interviews and recorded group inservice sessions. A constant comparison method was used, through triangulation of data, to determine the emerging assertions arising from the study.

Results of Study One

While the two teachers raised many issues over the course of the program, the following three areas were identified as central to their concerns: effective assessment of student learning; appropriate setting up of writing tasks; and the teachers' perception of their changing roles. These are reported in a series of assertions related to each concern.

> Teacher Issue 1: The teachers viewed writing in science as primarily an excellent technique for on-going assessment.

The views of the two teachers on the relationships between student writing, learning, assessment, and classroom program design, changed during the course of the in-service. At the outset of the program both Chris and Alan tended to see the main purpose of writing as testing students' knowledge after learning. However, as a broad range of writing types and purposes were trialed in class, both teachers perceived that student writing could serve many different assessment purposes and types. Both noted that student writing could be used to ascertain students' initial or emerging understandings rather than simply to grade students' knowledge at the end of topics. Chris commented in year one that:

> Sometimes when you are reading ... you really do see that they have misunderstood something and concepts they have got, perhaps, might be incorrect. You do get more surprises that way. I think when you gloss over a test sometimes it doesn't stand out as clearly ... it doesn't seem to pick up the concepts ... if you get so many right and so many wrong in a test well you get a final score and I guess you would assume they have good, bad or intermediate understandings. I found with the essays in particular you really do get an understanding of how much they have remembered or learnt or what concepts they have, so its good from that point of view.

Chris also found initial student writing on a topic, in the form of a concept map or brainstorming, provided a valuable indicator for subsequent unit planning. For example, he stated that students often "have a very large knowledge base" for some topics and therefore aspects of topics need not be covered in class. Alan agreed with this assessment purpose for student work, and believed that students' writing "allowed you to see their misconceptions, so you can work on them" (Alan, year 2). Both claimed that student writing gave them an accurate picture of student thinking and its justification. As Alan noted in year two:

> I now tell the students they've got to justify any answer they give me, so I think that's an important thing, that I just don't say to them yes/no, that's wrong and leave it at that. When they give an answer I want to know why it's wrong and why they've said a particular thing.

> Teacher Issue 2: The teachers required scoring criteria for assessing written products.

The assessment issue entailed various concerns and changing attitudes about different purposes for assessment, the teacher skills required to assess non-traditional writing, and what should count as evidence of student learning in and through writing. At the outset of the program all the teachers, including Chris and Alan, were concerned that they lacked expertise in assessing written work where assessment did not focus on correct terminology and concept knowledge in traditional short answers, tests, or assigned projects. They tended to see scientific content, linguistic expression, and student creativity as distinct separate categories. As Alan commented at the outset of the program, "student writing was difficult to correct because I don't know whether I should be correcting students' expression and spelling as well as the science concepts that they do or don't understand." Chris initially shared this view that 'creative writing' was difficult to evaluate because it added further 'new' elements to assessment. Extended discussion with the group led to a consensus that non-traditional writing was not "creative" in the sense that students were free to formulate fanciful explanations, but rather that student 'creativity' consisted in attempts to link, define, explore or translate scientific concepts in convincing and coherent ways through metaphor, analogy, narrative, or various comparisons.

Both teachers recognized that the requirement that students should explain or represent science concepts in different kinds of writing entailed additional writing demands compared to traditional tasks. Chris further noted that students who were very weak at writing generally were then doubly disadvantaged in trying to produce extended texts explaining science concepts. He addressed this concern by encouraging these students to represent their understandings in pictorial or verbal form. The teachers' general concern about distinguishing between students' scientific content knowledge, language skills, and 'creativity' formed the basis for continuous analysis of many examples of students' work through the first year of the program. At this time Chris was concerned that student writing was often well written and imaginative, but lacked a demonstrated understanding of science concepts. This issue was partially resolved through the development of assessment grids that specified target concepts that needed to be addressed, as well as the relative assessment weighting given to science concepts (80%), expression (10%), and creative cohesion of the text (10%). These grids also functioned as criteria sheets that were distributed to students when written work was assigned. However, by the end of the in-service program and many evaluation sessions both Chris and Alan recognized that it was difficult to separate assessment of concepts and assessment of student language, because as Chris commented, "the words students use show you how much they actually understand something."

> Teacher Issue 3: The teachers were concerned about identifying in-depth understanding from student written language.

The issue of what should count as conclusive evidence of student learning in writing remained an ongoing concern throughout the course of the program. Initially most of the teachers believed that appropriate student use of technical vocabulary and 'correct' or authorized definitions was reliable evidence of learning. By contrast, Chris argued from the outset that a student repeating the correct 'label' did not prove that the student understood the nature of the underlying concept or that the student could link concepts in a meaningful way. Throughout the program he urged the other teachers to look for extended or elaborated explanation of terms, or the ability of the student to connect concepts in the writing, concurring with Sutton (1992) that this was a more reliable guide to deeper student learning. While both Alan and Chris, like the other teachers, were concerned that students should know the authorized vocabulary and underlying concepts of particular topics, Chris argued that all too often students' mastery of scientific terms masked their lack of understanding. As the teachers became more familiar with diversified writing types, they generally recognized that these extended student texts enabled them to make far more precise judgments about the extent of, and gaps in, student knowledge.

> Teacher Issue 4: The teachers sought to gain understanding of how much to guide or direct student writing.

All the teachers in the program had many initial concerns about planning, writing task design, scaffolding for student writing, and the development of manageable classroom routines. Alan spoke for the whole group when he expressed an initial concern about the need for "a structure to allow the learning to take place." This need was met from the outset through collegial design of tasks, and the use of writing templates to provide frameworks for student writing (Prain, Hand & Kay, 1996). These templates were not as detailed as the linguistic guidelines outlined by some other researchers (Unsworth, 1997; Scheppegrell, 1998) on the grounds that such templates needed to strike a balance between giving students sufficient procedural and structural knowledge to guide their writing, but not so much advice that the writing task was turned into a rote activity, as Chris said, of 'filling the blanks'.

Both Chris and Alan believed that it was important for students to develop a strong sense of grappling with understanding concepts through writing and to feel a sense of ownership of their writing, effects that too much textual prescription undermined. Examples of other students' writing were also used to provide students with guidance. Chris noted that "if they have examples they will do a lot better job in most cases than if they have no idea where they are heading: they need direction." However, Chris considered the choice of example as crucial in that some students would attempt to reproduce the content of examples, therefore, he presented examples of different topics or ones that focused on different science concepts to those students had to address. For example, if students were asked to write a creative piece about heat transfer, Chris presented them with an example on red blood cells. In this way the writing process fitted both the 'forward search' and 'backward search' models of composing as outlined by Klein (1999) in that writers were expected to address a global goal, but also use the emerging text to clarify, explore or translate scientific concepts into non-technical language.

> Teacher Issue 5: The teachers found that creating authentic readerships, modeling writing, and brainstorming discussions were effective writing to learn scaffolds.

Both Chris and Alan also recognized through reflection on the early stages of the program that many students needed further specific scaffolding of writing tasks (see Hand & Prain, 1995). This led to specific target concepts for writing tasks to be nominated, and opportunities built into the classroom program for students to confer in groups and with the teacher about the adequacy of their emerging drafts to explain or link particular concepts. Further scaffolding was also provided for students on how to understand and address the issue of writing in science as communication. Chris noted that his younger students had little sense of how their choice of language changed, or should change, for different readerships. Working with an English teacher in the school he devised a task to help students become more familiar with the demands of writing for different audiences. He asked students to explore an incident in which they got into trouble at school. They were asked to explain themselves in writing to the year coordinator, their parents, and to friends. When this writing was read out in class students saw how the language and the focus changed, as well as their skill in evidence selection for different audiences. He explained that the exercise was intended to "get them [students] to understand the whole thing of how a different situation could be seen through the same person's eyes depending on what group they were talking to."

This skill in meeting the expectations and knowledge of different readers was then put into practice with students writing to communicate about science for different readerships. For instance, Chris had a Year 7 class write travel posters advertising space travel for the public, telling them that these posters were to be displayed at a local bank, with paper provided for public feedback. Alan's year 10 class produced brochures for farmers on the dangers of blue-green algae and invited a small group of local farmers to review their final product. Alan also had students write for their peers in Year 10 on the topic of motion, incorporating supportive material such as cartoon strips and pictorial graphs. Both teachers saw the provision of authentic audiences and purposes for student writing as further ways to provide effective contextual scaffolding for student efforts.

To provide further understanding of the links between writing tasks, science concepts, and successful communication with different readerships, both the teachers invited the researchers to teach their respective classes, to model guided discussion and set up group work to explore these issues. As they gained experience from these sessions, they used a variety of strategies to focus student thinking and support different stages of student writing. As noted already, this entailed various strategies, including use of templates to prompt student thinking (Bereiter & Scardamalia, 1987), student re-drafting of computer-based texts (Prain & Hand, 1996), detailed criteria for writing assessment distributed to the students when work was assigned, and the use of various student-generated diagrams, captions, and 'textbook' explanations as models for further student responses. Both Chris and Alan commented on the crucial importance of discussion as a clearing-house for ideas and as a way to develop and refine student focus and interest. For example, Chris commented that classroom discussion of relevant conceptual knowledge for a topic was a very important element in guiding students' writing. He explained that often during these extended discussions "kids came up with virtually a white board full of stuff." He saw this preliminary work as crucial in enabling students to identify strengths and gaps in their understanding, and to give them starting points for writing.

> Teacher Issue 6: The teachers became more aware of the value of writing-to-learn for constructing conceptual understanding over time.

Both teachers' understanding of how writing served learning changed and developed over the course of the program. Initially they viewed writing as mainly a record of past learning. However, as they become more comfortable with assessing various kinds and purposes of student writing, they began to see writing as a resource for facilitating conceptual development. Alan noted that it was important to structure writing tasks, whether using templates, class discussions, past models of writing, or other prompts, in such a way that the students had to develop their own individual interpretations of concepts rather than providing too much teacher input. In fact, he sees overload of teacher information as a "danger," because it could supplant the value of the writing as a real space for concept clarification and linkage. Chris commented that "it [writing] allows them to see their own construction," concurring with Klein's 'forward search' theory of how writing serves learning by enabling the writer to inspect and review emerging understandings. Chris also observed in year 2 that writing allows students "to see a sense of progress and learn things along the way rather than display ideas." He also thought that writing enabled students to "put all those concepts together". In elaborating on this comment Chris claimed these diversified writing tasks required the students to "operate at a second order level", that they have to embed what they into some other type of writing. In other words, "you have to not only know it but then be able to transfer it" (year 2). Both teachers recognized that students also needed to be able to read and write in the traditional genres of classroom science, but that diversified writing forms provide effective additional pathways to learning. They claimed that there were learning gains, in terms of student attitudes and levels of task engagement, from broadening the range of student writing tasks. These accounts suggest that both teachers believed that writing supported learning by enabling learners to explore, clarify, sort, review, elaborate on and organize their understanding of concepts and conceptual linkages within topics. Their views support the explanatory claims made for the four theories of how writing serves learning identified by Klein (1999).

> Teacher Issue 7: The teachers realized the need for their own roles to change for students to gain maximum benefit from using writing-to-learn strategies.

Both teachers believed that the program encouraged them to move from a self-perception as wisdom-giver to facilitator, although, as Alan noted in year two, "the circumstances in the classroom at different times causes you to take on different roles". Chris described himself as sometimes having to adopt a "lion-tamer" persona even though he would prefer to be a tour guide: "I'll direct them to a place and ... the people I'm guiding can then investigate themselves." He viewed the issue of increasing student ownership of tasks, including choices about particular categories of text type or audience, as an important element in enhancing student engagement. Alan also recognized the learning benefits associated with increasing student control over tasks, suggesting that "as you hand over control to the students I think they enjoy the classes more and you do get more out of them."

However, while Chris was committed to increasing student choice over tasks, he still preferred to frame the learning context. He claimed, "I'd like the kids to own it ... although I like to manage how they own it. I still believe there is no point unless they own it themselves." He believed his role was "to manage the way in which they get to learn, how they take it on board, what they take on board, how they represent it. I think it's my job to manage that." Alan also appeared to be comfortable in handing over greater control to students providing the task was "framed within a structured context." However, in other statements, he recognized that some students found decision-making difficult, because "sometimes too much choice is too much for some students to handle and they get confused with what to do." Both teachers recognized that guiding effective writing to learn in the classroom required them to play many varied roles within a learner-supportive environment, including personal tutor, sounding board for ideas, discussion leader, critical reader, editor, instructor, and feigned ingenuous reader to encourage students to elaborate their understandings of concepts.

STUDY TWO

Context and Methods

The research was conducted over two years (1999-2001) with three teachers and their students in Year 7 (11-12 year olds) and Year 10 classes (15-16 year olds) in two regional secondary schools in Iowa, U. S. A. Data were collected through semi-structured interviews, teachers' journals, field-note records, classroom observations and recorded group interviews. Semi-structured interviews with teachers focused on such issues as the changes in teaching roles required to implement the SWH, the necessary planning and support structures needed, teachers' perceptions of the SWH initially, during the study, and at the completion of the research, perceptions of student engagement with the new approach, and directions for future use of the SWH. The teachers met with the researchers to plan, implement, and evaluate the introduction of the SWH over six inservice meetings each year. Each of these meetings was tape-recorded, with discussion focused on the type of students who had completed the work, the involvement of these students in science generally, the value gained for the students and the teacher, their impressions of the effects of the SWH, and the reaction of students to the changes from usual classroom routines. Two research assistants made classroom observations and recorded field notes. Teacher interviews, group interviews, and field notes were coded and analyzed using constant comparison analysis in order to identify recurrent themes and generate assertions.

The two male teachers involved, James and Gary, were experienced secondary science teachers, with over 25 and 15 years experience of teaching science respectively, and Linda was a beginning teacher taught who had taught science for one year. Each teacher participated in a post graduate course prior to the study that focused on implementing student-oriented teaching/learning approaches to secondary science, and each teacher became familiar with the SWH during their study. The teachers had several opportunities to trial the SWH, however this report focuses on their use of the SWH to teach the topic of cells in biology for Years 7 and 10. For this topic, the following teaching concepts were identified: living things are made up of cells according to the cell theory of life; varying levels of organization and specialization exist among cells; and structure influences the functions of cells. Each teacher had experience over the two years of conducting at least eight SWH reports on different concepts relating to cells.

Results of Study Two

There was broad general alignment in terms of the teachers' concerns, although at the end of the study each teacher suggested different future issues and contrasting specific ways to modify the SWH to suit their individual teaching preferences. The findings have been grouped into five assertions, with the first three addressing initial concerns and the last three addressing further issues.

> Assertion 1: The teachers were concerned about the ability of students to generate the cognitive and metacognitive skills necessary for completing the SWH.

The three teachers expressed various concerns about the demands that the SWH made on students' knowledge and skills at each stage of engagement with the template. The teachers were particularly concerned about the students' ability to frame worthwhile questions at the outset of the topic. Gary noted that "students did ask good questions, but their questions were broad and difficult to answer through a lab." He considered that they had "little prior knowledge of cells" and therefore "were not able to ask more specific questions". While he considered that the students "really enjoyed asking questions", he considered these questions not very useful for framing investigation in the laboratory work. He believed from the outset of participation in the research study that guiding "pre-lab discussion" was vital to the success of the SWH. James commented that when he first attempted to implement student choice of questions, the students were "all over the map", where most of their questions were "meaningless and trivial". He noted that many students would ask "just yes/no questions" and "saw no problem with that." However, when he took control of the questions "the students lost their investment in the meaning of it." He commented that sometimes there were learning gains for students when they were forced to discover over time the ineffectiveness of their initial questions, "this is going nowhere," and then had to negotiate new questions. Despite this approach being viewed as "time-intensive", James considered this approach as "worth it" in terms of learning outcomes and student commitment to inquiry. He commented:

> You watch kids that would put out questions that were valuable questions but there is no way that they are going to be able to find an answer, but they actually produce a better learning situation than the kids that took a simplistic question - I can get this, I can do a good job, yes/no question, I'll get done, easy, quick, accurate.

James considered that students gained a valuable learning experience through identifying what represented a worthwhile question, even if this process took additional time. Linda was also concerned about the problems associated with giving students too much freedom in choosing questions. She noted that "it's like we are caught in a delicate system here because we are asking them to be free in their thinking. We tell them you choose, you have all the control, we want you to do this but then really what do we get out of it?" She considered that the activities need to be structured "because the tools that we have available for them to address these questions are limited." She believed that the students should be made more aware of the resources required for particular types of investigations. She also saw the limitations of the students' questions as indicative of their lack of prior knowledge of cells. She assumed that "most students would have a mental framework on cells on which to build their understanding", but then considered that "building a knowledge base" was a necessary pre-condition to eliciting worthwhile questions from the class.

The teachers also identified that students had difficulty with other demands of the SWH. They commented that many students struggled to distinguish evidence from claims, or demonstrate an understanding of scientific reasoning processes. James noted that working through the SWH with individual students often entailed a "step by step algorithm", where you had to focus very carefully on one claim at a time compared to the usual method of "traditional lab write-up." As he commented, "I can't remember how many times in those step by step processes we would turn back saying this has no connection to your question." All the teachers commented that the students struggled to identify the component parts of an adequate scientific explanation, suggesting that this kind of analysis was not required or addressed in students' past work on laboratory reports. In commenting generally on the students' role in learning James noted that many students wanted to play the role of "passive receptors as opposed to the directors of their own education and learning." He considered that when engaged in the SWH, students are "masters of their own knowledge and how to understand it ... which is really at odds with what they think or have been trained to think as far as what they are as students."

> Assertion 2: The teachers were concerned about how much scaffolding should be provided at various stages of implementation and struggled to find the right balance between direction and open-endedness.

Each teacher considered that effective implementation of the SWH represented significant demands on their skills as teachers and posed a range of challenges, especially in their first attempts to use the template. James, the most experienced teacher in this study, claimed that the SWH made more demands on the teacher in that he or she had to know when to give students "the support they need, and know how much to give when in the process". However, he considered that if the SWH was set up effectively, then students would provide "mutual support" where they scaffolded strategies for one another. Linda noted that "the dilemma for me as the teacher is how much control do I exert and when do I exert control over the process." She considered in the first instance where students were unfamiliar with the SWH, then it was necessary to provide questions and structure, but that "as students became better at the SWH then I could begin giving more control to the students." In the early implementation stage she specified the questions and tests for the topic, and expected students "focus on claims, evidence, reading and reflection". She believed that "some of the students' questions were so simplistic that they couldn't even see why you needed evidence."

Gary believed that implementing the SWH made significant new demands on his planning, his teaching style, and his normal classroom practices. He considered that it was crucial in using the SWH to frame the right questions concerning laboratory work. He noted that "in the beginning in helping students learn the process I think it is vital for the teacher to have a tighter structure to the process", because this "allows students to focus on the content to be learnt and not as much on the question and test." Because he considered his students' backgrounds to be "limited in asking a question and developing a test or experiment, he saw his role 'at least for the first 2 or 3 SWH labs' as 'giving a tighter structure to the labs." However, like Linda, he believed that with more experience of the process he could "give more and more control to the students." He was also concerned about the SWH's requirement that students should be given sufficient time to attempt to formulate their own explanations including evidence for their viewpoint. As he said,

> The hard part was letting the students think without my intrusion. If they gave an inaccurate idea I had to be silent and not rephrase or say, "No, you're wrong." I also am troubled by the amount of time it took to implement this part. I think in the future when we are familiar with this process I won't take a whole class period for this part.

Assertion 3: The teachers indicated that implementing the SWH required greater time than using traditional laboratory reports.

Each teacher perceived that effective implementation of the SWH took far more class-time than coverage of topics with more traditional activities. They also noted that students needed at least three exposures to the process of tackling an SWH before they developed any strong understanding of its purposes or value in terms of learning, and that students needed to be familiar with the stages and purposes of the SWH. As already noted by James, some students struggled to cope with the demands of developing convincing links between claims and evidence. James commented that it took one student "seven SWHs before she finally figured out that there was a link between claims, evidence, and observation, because they were all the same things in her mind…it just took a long time." However, he also considered that the value of student exposure to the demands of the SWH far outweighed the necessary time requirements. Gary also commented that the SWH entailed much slower progress in terms of topic coverage:

> I am also battling the feeling that there is so much material that has to be covered and that at the rate we have been going we are taking too much time and I want to take a day to lecture and revert back to the ease of this method of dispensing knowledge.

Assertion 4: The teachers invented their own strategies for promoting the exploring, questioning, and flexible thinking that is required for students using the SWH.

The three teachers agreed that the SWH provided a flexible approach to laboratory work and enabled them to gain deeper insights into the students' thinking and reasoning. They also indicated that they had made, or considered making, various changes to the format to suit their own teaching styles and perceived learning opportunities for students. Gary commented that in the second year of the study he provided set questions because "the kids came up with questions that were too difficult to answer." He noted that sometimes in the first year the students realized that their own question was inappropriate, so he permitted them to rephrase his question:

> So I allowed them to make those changes because it's a part of the process reflecting back. No that it isn't the best way to ask it. And the kids have that option. It allows them have some of that freedom to make those changes. But I kind of dictated at the beginning what they are going to do. But I always left them open. You can change the question if you want to if you find out you don't like the question, you want to ask it differently. That's okay.

Gary claimed that often in practice students are exploring and questioning on their own "and the teacher is guiding them through that by posing more questions." He noted that he used a lot of questions, but the students "are not being told what they are to do." He commented that he did not allow them to think "what the book says is always right," noting that "they have found instances when the book is wrong or there is a disagreement about ideas." In this way he claimed that his use of the SWH was "definitely very inquiry- oriented, but the teacher has to be the key in that." He also pointed out that there was a danger that the SWH template questions could be treated too rigidly and therefore turned into a "cookbook" where students would "not get much learning out of it."

Linda believed that individual teachers could implement the SWH effectively in a range of ways "as long as they stay flexible in their thinking."

In the second year of using the SWH, she increasingly saw the need to activate prior knowledge before asking students to formulate questions. She considered that students gained by getting "a little taste of what they are going to be able to do." For example she would ask them to look at red blood cells underneath the microscope, while not telling them which ones were red blood cells and which were white blood cells. Instead, she wanted the students to become engaged in the topic, so that they might start to formulate better questions based on this preliminary experience. However, she also noted that once students were addressing their questions, they sometimes "get lost in their data… in that absorption in what they are seeing." She considered that many students "forget about their question and can't even conceptualize how to answer it, and if it is answerable by what they are looking at." She saw her role as giving the students "that taste and then they can go back and say now what would be a good question to ask about what you can see."

All three teachers mentioned how students need to adjust their thinking to become flexible and authentic thinkers while using the SWH. James commented that the SWH encouraged students to "invest some of themselves into the process," especially in relation to questions or claims or evidence. The students "have to really build an argument and see if that argument will stand in comparison to classmates and their work". He thought that the SWH can "promote thinking and can produce thinkers" because "if you stay with the cookbook you produce technicians…. you have got people that can follow a cookbook recipe and crank it out without an iota of thought." He saw a major gain of the SWH was that students "now accepted ownership" of their reasoning, that it is "up to them and they can build it if they want to, but nobody else can do that for them." He claimed that the idea of claims and evidence really promoted thinking because students had to put their ideas "out there and then they are going to have to defend them or they will get slaughtered if they don't have a connection between claims and evidence. That is so hand in glove with what you try to do in a classroom."

James saw value in students building and defending their ideas, "A lot of them even quit asking what's the right answer. The idea is here's what I think, why do I think that, and then be able to defend your thoughts. But boy is that foreign, it's like pulling teeth to get kids to have their own thoughts and defend them." Gary also believed that the SWH enabled the students to "deepen their thinking on a particular topic or concept" because they have the "freedom to articulate their thinking with peers." He viewed the SWH as more flexible than traditional reports where "everything had to fit some place."

CONCLUSION

The assertions generations from these two studies of teaching writing to learn in science indicate that teachers believe writing is conducive for constructing authentic, flexible, and deep understanding of science concepts. However, teachers in both the studies recognized that these gains come only with increased time and hard work in terms of creating new student and teacher roles. In common from the two studies is the assertion that teachers struggled with when to give guidance and when to leave cognitive decisions to the students. The implication is that teaching writing to learn will cause teachers to continually balance teacher and student cognitive activities.

Both the studies indicate that teacher implementation of new strategies is not easily achieved and requires that teachers concerns be met. All the teachers involved in the two studies had the advantage of being involved in long term professional development activities where they were supported in implementing writing to learn strategies. This is not always the case when teachers attempt to use new strategies. Importantly the teachers in the study identified that there were major changes that were required in the pedagogy to be used, the role of the students, and their own roles. There was great importance placed on ensuring that students were given opportunities to think and display their thinking, which all the teachers believed was beneficial for developing their understandings of the required pedagogy. By adopting strategies which were much more student-centered the teachers were aware of the need to address students' ideas, plan ways to use these ideas in terms of the activities, promote more student self awareness of their own thinking and construct assessments to reflect the deeper level of thinking achieved by the students. Such change is difficult without support via such mechanisms as the professional development activities described.

The exciting part of the two studies was the teachers' discussions about their own individual strategies for implementing writing. When teachers have reached the stage of owning the new strategies then they are more prepared to move past technical implementation and become more facilitative in their pedagogy. Until this stage is reached, then the rich learning environments required for maximum gains are very difficult to achieve. However, recognition must be given to the time required to achieve this stage. These two programs were extended in terms of the commitment of time and energy.

CHAPTER 9

BREAKTHROUGHS, CLASSROOM IMPLICATIONS, ON-GOING, AND FUTURE RESEARCH

Carolyn S. Wallace, Brian Hand, & Vaughan Prain

Taken as a whole, our studies provide data that support writing as one mode of science learning. Our research has gleaned some new understandings about how students view the mechanisms of writing to learn, and some of the types of tasks and conditions that promote learning science through writing. In this chapter, we will describe both what we consider to be breakthroughs from our work, as well as areas of research still untapped. We will also address classroom implications of the research.

BREAKTHROUGHS

Writing and Conceptual Understanding

Research in this volume, including both the authors' research and the studies outlined in Chapter 2, continue to provide evidence that writing to learn does indeed enhance students' conceptual understanding of science. The theory that writing supports conceptual knowledge building in science is now well documented. Results of the quantitative studies reported in Chapters 5 and 6 of this volume show that writing is efficacious for promoting better performance on high level conceptual test questions, as well as, multiple choice questions. There is qualitative evidence supplied in Chapter 6 that using the SWH promotes conceptual thinking and conceptual change in some cases. These studies enhance previous evidence, such as that generated by Mason and Boscolo (2000, 2002) and McCrindle and Christensen (1995), that writing promotes deep conceptual understanding for students.

Like any instructional intervention, not all students may benefit or benefit to the same degree from any one experience of writing in science. As originally proposed by Langer and Applebee (1987), the intellectual benefits derived from writing to learn are very sensitive to the tasks the children are asked to do, aspects of the educational environment, such as the teachers' style of instruction, students' past experience with writing, and disposition towards using writing as a mode of learning. Even with these caveats, we believe that it would be well worth teachers' and students' time to use writing regularly as an instructional tool to build understandings of both science concepts and the nature of science. Our studies indicate that the use of diversified language in science, including rewording scientific text into everyday language is a beneficial activity for students. Future research is needed on more refined research questions of the effects of tasks, science topics, environments, and the responses of more diverse students, rather than the basic question of whether writing supports learning.

Socio-cognitive Mechanisms

Studies in this volume have informed us not only that writing does increase conceptual understanding, but also something about the socio-cognitive mechanisms through which that change in understanding might take place. The mechanism most alluded to by the students in our interview studies might best be captured by the theme, *turning others' science into "my" science.* Writing to learn strategies tend to emphasize personal expression. This seminal idea was put forth by Britton in 1970 and our data appear to support it strongly. It is the power of personal expression, linking students' known conceptual categories and language to the target language, that seems to be key for students to engage cognitively and metacognitively with the writing task. When engaged in this linkage, students come to view writing as an authentic expression of what it is they know.

Authentic meaning negotiation and language use may be viewed as a key to learning from writing in science. Our data support this as follows. First, students expressed to us (as shown in Chapter 7) that the writing tasks they found most useful included framing their own questions, explaining a process, and rewording science content for a different audience. In each of these tasks, students are given both the freedom and the responsibility to generate meaning in their writing. Framing one's own questions for learning requires personal commitment, as well as, deep scientific thought. Students necessarily commit to learning the answer to the question, when they pose the question themselves. Explaining a process similarly requires a personal commitment to get at the heart of scientific meaning; to explore what it is that one knows and does not know. Rewording science content for a different audience also implies personal choice about language use that will be viewed by the audience. To engage authentically with this task, students must make a commitment to attempt language that will authentically express their best understanding. When students engage with these tasks, they therefore commit to meaningful learning and the transformation to personally understandable forms of science content.

The second set of data in this volume that support the theme of *turning others' science into my science* as a socio-cognitive mechanism for learning from writing is that on the construction of question, claim, and evidence. Our studies indicate that the use of the SWH has been strikingly useful for generating these understandings. The data in Chapter 6 indicate that when using the SWH, students come to understand these constructs, how they are created and used in scientific writing, as well as their connections with each other and with laboratory investigations. As Bazerman suggests (personal communication, September, 2002), inverting the voice from "someone else" to the students' own voice (What can I claim? Why am I claiming this? What is the evidence for this claim?) fosters the generation of personal meaning for the laboratory investigation. Students may become authentically engaged when it is they who are making a claim and backing a claim with evidence. This kind of first person engagement with the task can foster understanding of the nature of science. The inversion of the scientist's voice from other to self helps the student to make meaning for laboratory investigations that heretofore have been inauthentic. Thus, writing to learn is one way in which laboratory can become a meaningful science experience.

Metacognitive Mechanisms

The studies in this volume support and add information to previous studies indicating that metacognition is a key to understanding writing to learn. We have learned from our interviews that students recognize writing as promoting deeper thinking, making connections, and seeing gaps in one's own knowledge. As described in Chapters 6 and 7, and as theorized (Emig, 1977; Klein, 1999; Bereiter & Scardamalia, 1987), since writing requires formulation of ideas in language, it necessitates evaluation of the content that one has ready to express. One of our studies, the South Hall erosion study described in Chapter 6, provided evidence that the backward search mechanism, that of setting subgoals for writing and then searching for scientific hypotheses and evidence to reach those goals is accomplished by students as young as eighth grade. Student interview data in Chapter 7 indicates for high school chemistry students, discovering gaps in their own knowledge of stoichiometry was a powerful metacognitive tool for generating new understandings. This implies that a forward search mechanism also comes into play in writing- as students wrote and reviewed, they discovered gaps in their own knowledge that promoted them to fill in the missing information.

A forward search mechanism is also supported by our findings that an initial planning experience is more valuable than a delayed planning experience. Planning ahead of writing allowed students to see what additional content was necessary to generate.

There is also evidence in our studies for the genre knowledge hypothesis of writing to learn (Klein, 1999). The genre hypothesis asserts that students learn science as they work to fulfill the demands of a specific writing genre. In our studies, the SWH provided a genre scaffold for negotiating meaning from laboratory investigations. As students became familiar with the SWH template, they learned the scientific genre for writing a question, claim, and evidence. Thus, they built an understanding of these genres and in turn, enhanced their understanding of the nature of science. On some occasions, the quest for evidence also led to the building of greater scientific conceptual understandings. As one of our interviewees stated, "when you are writing, you have to know what you are talking about, you can't fake it." In this way, the SWH fosters the learning of epistemology of scientific investigation, and how arguments in science are constructed.

Finally, students have informed us that *language transformation* is an important component of writing to learn. Writing tasks that provide opportunities for transforming language into an authentic voice is a powerful tool for generating the metacognitive mechanisms that foster science learning. When students have to negotiate meaning in every day language and/or transform meaning into a more scientific genre, such as claim and evidence, they necessarily negotiate meaning. Students in our studies enjoyed reframing explanations for authentic audiences including parents, peers, and younger students. They recognized the amplification of meaning that such a language transformation provides. Similarly, the negotiation of meaning for laboratory investigations done in peer groups helped students transform language from every day notions into scientific genre. Thus, we assert that negotiating among multiple discourse types, from more scientific to every day and vice versa is key to stimulating metacognition and learning from writing in science.

CLASSROOM IMPLICATIONS

The results of our studies indicate that writing to learn in science activities will be appropriate in any classroom where authentic conceptual understanding is a goal of learning. The time it takes to implement writing to learn is well offset by the conceptual gains that occur. In addition, results of some of our studies indicate that writing to learn is effective for preparing students for multiple-choice tests, a result we did not anticipate. Therefore, writing may help students with preparing for high stakes tests, such as state required graduation tests. We believe that our research has helped pinpoint some specific implications for classroom teachers considering the types of writing that students should be asked to engage in.

First, activities that involve a transformation or elaboration of language for a different audience are of high value. Teachers may assign students writing tasks that transform science content from other print-based media, or from their own lab investigations to a new authentic piece. Students may write for peer students, younger students, their parents, a governmental agency, or the general public. They may write content in terms of letters, textbook sections, tradebooks, magazine articles, brochures, or scientific posters. The key is to engage students in translating scientific terminology into everyday language.

Second, planning ahead of a first draft will assist students to engage metacognitively with a writing task and search for additional content. This may lead to enhanced conceptual understanding from the writing task. We suggest that teachers involve students in small group or whole class planning when given a writing assignment. The setting of goals and subgoals for writing may evoke both a backward search mechanism in which students search for content to obtain their writing goals and a forward search mechanism in which students review what they have planned and new connections to content are stimulated. We have also observed that receiving feedback on a first draft from peers or the intended audience is very useful to students, although we haven't formally researched the efficacy of feedback. When possible, teachers can facilitate the feedback from audiences such as younger students, or students in an English class.

Third, our studies that focus closely on laboratory investigations (those involving the SWH) indicate that students do indeed engage more deeply when they are allowed to pose their own questions for investigation. They can be guided by whole class discussion and suggestions from the teacher, but then allowed to reformulate the question in a way that makes sense to them. Because scientific research is typified by a chain of reasoning from question to data to claim and evidence, establishing an authentic question that students understand is a cornerstone for an authentic experience of understanding. In our studies, students as young as grade seven enjoyed framing questions, and felt that they actually learned more when they were able to frame questions.

Fourth, the role of the teacher in promoting learning through writing is going to require an understanding of the changing pedagogy necessary for successful implementation. As the earlier discussions indicated, adopting writing to learn strategies does require a change in thinking and pedagogical practice. Teachers, as in accord with *National Science Education Standards*, need to focus much more on the conceptual ideas framing a topic rather than on the recall of the content within the topic. The use of writing to learn strategies encourages students to make the connections between the conceptual framework and the supporting content knowledge. Focusing on the conceptual frameworks for topics is not something that teachers in general are familiar with, and requires much time and effort. This is going to be a crucial dimension of the required pedagogy. A second and equally crucial dimension is that writing is a process of negotiating meaning and thus teachers will need to provide opportunities for students to negotiate meaning both at an individual level and a group level. Teachers will need to be prepared to allow, and encourage, negotiation for which there are multiple pathways for understanding to be developed. This does not mean that science knowledge is abandoned but rather students are given opportunities to negotiate meaning through writing to understand the concepts, rather than writing to provide demonstrate recall of content items. We know that students are individuals and thus will have different understandings and interpretations, which need to be explored.

ON-GOING RESEARCH

The elements of the SWH, which have been articulated earlier, focus student and teacher attention on engaging with the some of the practices of scientists. Building on the initial studies described, there are a number of theoretical issues that need to be more carefully examined. These issues need to be dealt with methodologically in order for us to begin to describe more robust generalizations and advocate a broad application of the use of the SWH.

Pedagogy

In recent papers delivered at an international conference on science literacy at Victoria in British Columbia, Canada, Jonathon Osborne and Rick Duschl argued separately for the need for scientific argumentation to become more central in the school science curriculum. Unfortunately, the concept of appropriate pedagogy was not adequately dealt with in these papers. However, the role of the teacher is critical in helping students to be aware of, and apply, scientific arguments well. While the student template of the SWH provides a framework that scaffolds the argumentation process, the teacher template clearly emphasizes the need for an active pedagogical role for the teacher.

The teacher's role is conceptualized as negotiating meaning with students. However, this immediately raises issues about what does this mean in terms of the questions required and the environment needed to have successful negotiations. While the final act for students is completing the written component of the SWH, the process of social negotiation requires that much discussion occur before this final act. What is going to be considered appropriate pedagogy and how can we measure the quality of implementation?

While there are a number of measurement instruments that are framed around observing teacher behaviors and patterns of questioning, we believe that a different focus on classroom dialogue needs to be examined. The SWH is based on providing opportunities for students to negotiate meaning, and thus we need to examine the nature of those negotiations. Clearly discourse analysis can be used to obtain a very rich analysis of the dialogue that occurs in an SWH classroom. However, we are suggesting that on a broader level, the directional nature of the discourse can be used a means of analysis to determine the level of pedagogical implementation occurring. Importantly for us, the concept of the discourse involved in negotiation means that there has to be a two-way dialogue involved. Negotiation cannot occur when there is unidirectional discourse or in other words, a monologue.

Our current classroom observations are centered on determining if there is an authentic dialogue or if teacher talk is essentially a monologue. We have been directing our observations at whole group and small group discussions because negotiation is expected at these levels. We are trying to determine if there is indeed communication occurring. Good communication involves both listening and questioning, thus we are trying to determine if the participants involved, either in whole or small group setting, are spring-boarding their conversations from the on-going discussions or if one dominant person is directing the conversation regardless of the input. We are not attempting to breakdown the conversations to analyze the discourse but rather trying to note whom is participating and is the conversation building as a consequence of the discussion. In small group observations we are observing the student discussion, and if the teacher joins the discussion, we are trying to determine if the nature of the conversation is changed.

From these observations, we are able to provide the teachers with feedback to assist them in developing their questioning strategies. This appears to be the greatest difficulty facing the teachers as they begin to change their practices. Teachers have been the gatekeepers of knowledge, and when we begin to implement the SWH, expecting the students to have to negotiate meaning in a more obvious manner, teachers can find this task difficult. Given that students are continually negotiating meaning regardless of what happens in the classroom, the importance of making teachers aware of the concept of negotiation is critical in beginning to understand the benefits of the SWH.

The function of observations is to not only focus on the nature of the dialogue but to begin to address concerns with the learning environment that is in place. We have placed emphasis on student-centered learning as a critical component of the process. However, this does not only refer to the nature of the dialogue. Our work with classroom observations is also helping teachers create a non-threatening learning environment. What are the crucial elements of a good learning environment that enables such negotiations to take place? In high school science in particular, the use of small group work that is focused on student negotiation of meaning has not been thoroughly researched.

We need to continually focus on the challenges facing teachers in changing pedagogical practices, adopting a focus that is much more on the concepts of a topic, rather than the content of the unit, and developing student reliance on having to negotiate and debate meaning. There is a need to further develop research tools and methods to assist teachers' change and implementation of the SWH.

Students' Perceptions

Using the SWH well means that students are required to take a more overt active role in the learning process. While actively constructing their own sense of meaning from any classroom activity, students using the SWH are now required to make that activity much more public than previously. Small group and whole class discussions require students to take an active voice not only in what is being said but also in negotiating the direction of the activities. From a research perspective, the question becomes, what do we need to observe and record to better understand the students' role and expectations when using the SWH?

There are two critical areas that we believe need to be addressed. First, there is a need to understand what students are doing when they are engaged with the decision-making and negotiating processes that occur as a function of group work. What becomes important to students within these processes? How do they function when given opportunities to have their decision making made much more public than previously? How do they deal with the science concepts and knowledge that are the focus of the investigations? Second there is a need to better understand how students engage with negotiating meaning when using the writing processes required when using the SWH. In particular, how do students perceive the need to have to make their own choices in terms of questions, claims, evidence and reflections, as a component of their learning? Does the scaffolding effect of the SWH help students negotiate meaning through the writing process? Data to help answer these questions will most like include close observation and/or video taping of students during group work, semi-structured interviews, and student questionnaires.

A new avenue of research that is beginning to show some promise is cognitive analysis of the text written by students. Coding texts to determine the types and frequencies of cognitive processes provides evidence about students' learning experiences during the SWH. Initial results from our studies on writing indicate that writing to learn strategies are helping all students, that is, the strategies do not disadvantage any group. However, an important question for us is, do low ability students use the same cognitive processes as high ability students when using the SWH? If the SWH scaffolds critical thinking and argumentation, can we analyze text to determine some of the cognitive strategies used by the writer? Our research group has begun to develop coding schemes according to the cognitive process that the text represents, that is, observation, method, claim, inference, etc. In this way, we may begin to reveal more about student thinking during use of the SWH.

Explanation and Argumentation

Another area of on-going research focuses on the connection of writing to explanations in science. While our work with the SWH has researched the construction of claims and evidence previously, we are also concerned with students linking their data and evidence with larger theories and explanations. One of our questions is thus, how can students integrate their authentic, negotiated knowledge within the larger explanatory framework of canonical science? We view the creation and/or application of explanations to be a critical component of inquiry-based learning.

In a recent study, we collected data from rural high school students in biology in which they wrote SWH style reports with an additional prompt to explain their investigation findings using all resources available to them. The preliminary data analyses indicate that students responded authentically to the task. That is, they wrote explanations that did arise from their negotiated meanings for the investigations, rather than parrot scientific language they found in their textbook. Our sense is that students will create explanations for their data that are very reasonable and consistent with their observations and claims. Whether these explanations mirror canonical science appears to depend on the students' epistemologies, prior knowledge, and comfort level for reading and interpreting authoritative information. Some students created their own unique examples and analogies to explain their findings, while others were comfortable appropriating more scientific discourse. Almost universally, however, students were not able to include all causative links in explanations requiring a sequence of causality (e. g. causes of sickle cell anemia). Much further work is needed to probe students' epistemologies and personal understandings of science in order to reveal the most useful curriculum and teaching strategies in inquiry-based, writing intensive science.

Currently, we have two projects on-going in university biology, one for a majors' biology course and one for a non-major's biology course. In the majors' course, we are implementing a special one-hour discussion session devoted to writing for learning in biology. Students in this course will be introduced to the topics of writing to learn, the nature of science writing, and tips for writing. The majority of the class time, however, is dedicated to discussion and peer critique of students' short written biology explanations (about 500 words). Along with the explanations, students are being asked to write an argument for a potentially more plausible explanation for the phenomenon under consideration, related to their evidence. Our research questions include the following:

1. What is the effect of writing explanations on learning biology concepts as measured by a course final exam in majors' biology?
2. How do explanations of biology students in a writing intervention change over time?
3. What is the evidence that biology learners' are utilizing any of Klein's (1999) proposed hypotheses for learning from writing: (a) point of utterance; (b) forward search; (c) backward search; or (d) genre knowledge?
4. How do students' epistemologies including nature of science understandings and learning beliefs relate to their use of writing for learning?

Writing as Assessment

In the non-majors biology project, we are adapting new assessment formats for the laboratory section of the course for the purpose of altering the course curriculum. Most students taking Biology 1103, Concepts in Biology, also enroll in a laboratory class of three hours per week. Many of these non-majors have little interest or background in science. The laboratory activities are currently outdated and in need of rewriting towards an inquiry-based orientation. The new laboratory assessments will be designed to promote: (a) student engagement and meaningful learning; (b) the inquiry-based nature of the course; (c) scientific literacy, including students' reading and writing skills in science; and (d) student achievement for incorporating meaningful biology concepts into the knowledge base. Changing the nature of the assessment is viewed as a vehicle to directly promote explicit learning objectives. The learning objectives we wish for students to achieve reflect high levels of scientific literacy including:

1. recognizing the connections among question, claim, evidence, and explanation in science;
2. designing and conducting an investigation;
3. writing a scientific argument, using claim, evidence, and warrant.
4. improving understanding of the "big ideas" in biology and how they relate to everyday life
5. improve critical reading skills

Our plan is to change the assessments in the laboratory section from several quizzes to a Scientist's Notebook, a Science Literacy Test, and two laboratory reports based on the Science Writing Heuristic (Keys, Hand, Prain, & Collins, 1999). Implementing these assessment criteria will facilitate the redesign of laboratories towards an inquiry-based approach, because to succeed on these assessment materials, students will need to understand the scientific constructs of question, claims, data, warrant, and evidence. They will be coached towards designing and conducting their own investigations, and thus, be able to write high quality laboratory reports, including a scientific argument to explain their findings. In addition, we propose that participating in inquiry-based laboratories and writing thoughtfully about them will transfer over to students' critical reading skills. We postulate that students involved in our intervention will score higher on a test of Scientific Literacy, which requires them to gauge the certainty of scientific statements, as well as identify core elements of any scientific study, such as the hypothesis, methods, and conclusions.

FUTURE RESEARCH

While research in writing to learn in science has progressed greatly over the last decade, many research questions remain. The most obvious gap in our database is the use of writing with diverse students. Studies of scientific literacy with students of non-White ethnic cultural backgrounds, special needs, English as a second language, lower tracked classes or urban or rural cultures are on the rise (see, for example, Varelas, Becker, Luster & Wenzel, 2002). However, research on the uses of diverse writing, such as those we advocate remains scant. While the research in this volume was conducted with largely heterogeneous classes, the majority of the students in the research were mainstream White children, both in the United States and Australia. Diverse children may have special needs related to writing instruction that have not yet been uncovered by research. For example, children with learning disabilities or English as a second language may find writing frustrating or exceedingly difficult and, therefore, be less able to benefit from writing to learn activities. Scaffolding instructional activities to assist children with writing difficulties could comprise an entire research program. Recent research in other modes of meaningful learning, for example, concept mapping (Fisher, Wandersee & Moody, 2000), could inform educators about bridges between different forms of language use. Conversely, writing about science in an authentic voice may appeal to some students traditionally underrepresented in science (Chinn & Hilgers, 2000). Thus, one large area of research in writing to learn is how and to what extent diverse children engage with writing to learn activities.

A second needed area of research deals with the assessment of writing to learn science. Our work with teachers indicated that they felt less confident in how to assess non-traditional types of writing in science, and there is a need to research the appropriate frameworks and professional development that would support teachers in assessing the various kinds of writing resulting from use of writing-for-learning strategies. Our work with teachers in designing appropriate topic tests for this learning also indicated that there are significant implications for assessment. We found that traditional short answer questions were not always effective in measuring students' learning, especially in the area of higher order knowledge and problem-solving. While the findings on student learning in our research were based on more diverse forms of assessment than short-answer questions, there is scope for further research into appropriate, viable assessment methods that enable learning gains from writing-for-learning programs to be measured accurately. Our research suggested that these learning gains were better measured where students have to demonstrate an ability to transfer science concepts to new contexts, and therefore assessment methods and tasks needed to be altered.

A third area of research that is little explored but has great potential is that of combining reading to learn instruction with writing to learn instruction. We know that students undergo the same interactive-constructivist process when reading that they do when writing (Holliday, Yore, & Alvermann, 1994). Thus, reading and writing may be mutually supportive cognitive/metacognitive activities. Research has indicated that even successful high school seniors have a great deal of difficulty interpreting a scientific argument in printed text (Norris & Phillips, 1994). One question, therefore, is whether experiences in writing, such as those encountered in using the SWH, could improve students' ability to read and understand scientific text. A second question is whether instruction in reading scientific text, such as that in the experimental report genre, could facilitate students' writing of a scientific argument. A third question is how reading can fill in gaps in student knowledge when writing a scientific argument or other type of text. For example, when conducting laboratory investigations, it may be desirable for students to move from claim and evidence to a more complete explanation or theory of the science involved in the investigation. While we believe that students can and should look for patterns in their own data and pose possible explanations, in some cases it is necessary to provide scientific theory as it has been constructed over time through the scientific cannon. In this case, reading could be profitably combined with laboratory investigation to provide a deeply connected experience between theory and lab.

A fourth area of needed research deals with the realm of technology. There is growing recognition that the use of new technologies for practicing and teaching science has broad implications for the role of writing in learning science. Researchers such as Jonassen and Carr (2000), Lemke (1998) and Linn (2000) have argued that these new information and communication technologies provide a platform of necessary processing and manipulative options, thus leaving learners free to engage in higher order problem-solving activities. There is a need for extended research into how writing for learning tasks can be integrated into these new ways of conducting and representing science activity and findings. Recent multimedia resources have also spawned new and hybrid genres for representing science concepts, methods and findings to an increasingly globalized audience. These text types include web-based, but print-dominant, genres for popularising advanced science, web pages, hypertexts with increasingly dense networks of links across meta-sites, simulations, interactive media and 3-dimensional visualizations. These developments pose further challenges for identifying effective student writing practices for learning.

Aikenhead, G.S., Jegede O., & Allen, N. (1999, March). *Cultures studies in science education: Students' indigenous cultures versus the culture of science.* Preconference session at the annual meeting for the National Association for Research in Science Teaching, Boston, MA.

Ainsworth, S. (1999). The functions of multiple representations. *Computers & Education, 33*, 131-152.

Airasian, P.W. & Walsh, M.E. (1997). Constructivist cautious. *Phi Delta Kappan, 78,* 444-49.

American Association for the Advancement of Science. (1993). *Benchmarks for science literacy.* New York: Oxford University Press.

Applebee, A. N. (1984). Writing and reasoning. *Review of Educational Research, 54,* 577-596.

Audet, R. H., Hickman, P., & Dobrynina, G. (1996). Learning logs: A classroom practice for enhancing scientific sense making. *Journal of Research in Science Teaching, 33,* 205-222.

Barnes, D. (1976). *From communication to curriculum.* Harmondsworth: Penguin.

Barnes, D. (1986). Language in the secondary classroom. In D. Barnes, J. Britton and M. Torbe (eds.), *Language, the learner and the school,* ,(3rd edition). Harmondsworth: Penguin.

Bazerman, C. (1985). *Shaping written knowledge: The genre and activity of the experimental article in science.* Madison, WI: University of Wisconsin Press.

Bell, B., & Gilbert, J. (1996). *Teacher development: A model from science education.* London: Falmer Press.

Bereiter, C., & Scardamalia, M. (1987). *The psychology of written composition.* Hillsdale, NJ: Lawrence Erlbaum Associates.

Berkenkotter, C., & Huckin, T. N. (1995). *Genre knowledge in disciplinary communication: Cognition/culture/power.* Hillsdale, NJ: Erlbaum.

Boscolo, P. & Mason, L. (2001). Writing to learn, writing to transfer. In P. Tynjala, L. Mason, & K. Lonka, (eds.) *Writing as a learning tool.* Amsterdam: Kluwer Press.

Brickhouse, N. (2001). Embodying science: A feminist perspective on learning. *Journal of Research in Science Teaching, 38,* 282-295.

Britton, J. (1970). *Language and learning.* New York: Penguin Books.

Britton, J. (1982). Shaping at the point of utterance. In Pradl, G. M. (Ed.), *Prospect and retrospect: Selected essays of James Britton* (pp. 139-145). Montclair, NJ: Boynton/Co. (Reprinted from A. Freedman and I. Pringle, (Eds.), *Reinventing the rhetorical tradition,* 1980, Conway, A.R., L & S Books., for the Canadian Council of Teachers of English).

Bruner, J. S. (1964). Course of cognitive growth. *American Psychologist,* 19 (1), 14.

Bruner, J. S. (1966). *Towards a theory of instruction*. Harvard University Press: Cambridge, MA.

Caldwell, B. J., & and Spinks, J. M. (1991). *Leading the self-managing school.* London: Falmer Press.

Campbell, B., Kaunda, L., Allie, S., Buffler, A. & Lubben, F. (2000). The communication of laboratory investigations by university entrants. *Journal of Research in Science Teaching, 37,* 839-853.

Champagne, A., & Kouba, V. (1999). Written products as performance measures. In J. Mintzes, J. Wandersee, & J. Novak (Eds.). *Assessing science understanding: A human constructivist view* (pp. 224-248). New York: Academic Press.

Chinn, P. & Hilgers, T. (2000). From corrector to collaborator: The range of instructor roles in writing-based natural and applied science classes. Journal of Research in Science Teaching, 37, 3-25.

Christie, F. (1981). The language development project. *English in Australia, 58*, 3-10.

Connally, P. & Villardi, T. (1989). *Writing to learn mathematics and science*. New York: Teachers' College Press.

Connor, W., Prain, V., & Hand, B. (1994). Improving learning through writing in secondary science: Two examples. *Australian Science Teachers Journal 40* (1), 42-47.

Crump, S J. (Ed). (1993). *School-centred leadership: Putting educational policy into practice.* South Melbourne, Australia: Nelson.

Darling-Hammond, L. (1998). Teacher learning that supports student learning. *Educational Leadership 55*(5), 6-11.

Department of Employment, Education and Training. (DEET, 1989) *Discipline Review of Teacher Education in Mathematics and Science* Canberra: Australian Government Publishing Service.

Dolin, J. (2001). Representational forms in physics. In D. Psillos, P. Kariotoglou, V. Tselfes, G. Bisdikian, G. Fassoulopoulos, E. Hatzikraniotis, E. Kallery (Eds.). *Science Educaion Research in the Knowledge-Based Society. Proceedings of the Third International Conference of the ESERA* (359-361). Thessaloniki, Greece: Aristotle University.

Donald, M. (1991). *Origins of the modern mind.* Cambridge, MA: Harvard University Press.

Driver, R. (1986, April). *Reconstructing the science curriculum: The approach of the children's learning in science project.* Paper prepared for the Annual Meeting of the American Educational Research Association, San Francisco.

Driver, R. (1990, September). *Constructivist approaches to science teaching.* Paper presented at the University of Georgia, Mathematics Education

Department as a contribution to the Seminar Series "Constructivism in Education".

Driver, R., Leach, J., Mortimer, E., & Scott, P. (1994). Constructing scientific knowledge in the classroom. Educational Researcher 23(7), 5-12.

Emig, J. (1977). Writing as a mode of learning. *College Composition and Communication, 28,* 122-128.

Erickson, F. (1986). Qualitative methods in research on teaching. In Wittrock, M. C., (Ed.), Handbook of research on teaching (3rd ed., pp. 119-161). New York: Macmillan Publishing Company.

Fellows, N. J. (1994). A window into thinking: Using student writing to understand conceptual change in science learning. *Journal of Research in Science Teaching, 31,* 985-101.

Fensham, P., Gunstone, R., & White, R. (1994). *The content of science. A constructivist approach to its teaching and learning.* London: Falmer Press.

Fisher, K. M., Wandersee, J. H., & Moody, D. E. (2000). Mapping biology knowledge. Dordrecht, The Netherlands: Kluwer Academic Publishers.

Flood, J., Jensen, D., & Squire, J. (1992). *Handbook of research on teaching the English language arts.* New York: MacMillan.

Florence, M. K., & Yore, L. (2002, April). *Learning to write like a scientist.* Paper presented at the annual meeting of the National Association for Research in Science Teaching, New Orleans, LA.

Flower, L., & Hayes, J. (1980). The cognition of discovery: Defining a rhetorical problem. *College Composition and Communication, 31,* 21-32.

Flower, L., & Hayes, J. (1984). Images, plans, and prose: The representation of meaning in writing. *Written Composition,1,* 120-160.

Ford, C., Yore, L. D., & Anthony, R. J. (1997). Reforms, visions, and standards: A cross-curricular view from elementary school perspective. *Resources in Education,* (ERIC), ED 406168.

Fulwiler, T. (1987). *The journal book.* Portsmouth, NH: Heinemann.

Fulwiler, T., & Young, A. (1982). Language connections: writing and reading across the curriculum. Urbana, IL: National Council of Teachers of English.

Galbraith, D. (1999). Writing as a knowledge-constituting process. In D. Galbraith & M. Torrance (Eds.). *Knowing what to write: Conceptual processes in text production* (pp. 139-159). Amsterdam: Amsterdam University Press.

Galbraith, D., & Rijlaarsdam, G. (1999). Effective strategies for the teaching and learning of writing. *Learning and Instruction, 9,* 93-108.

Galbraith, D., & Torrance, M. (1999). *Knowing what to write: Conceptual processes in text production.* Amsterdam: Amsterdam University Press.

Gee, J. P. (1990). *Social linguistics and literacies: Ideology in discourses.* Bristol, PA: Taylor and Francis.

Gee, J. P. (2001, August*). Language in the science classroom: Academic social languages as the heart of school based literacy.* Paper presented at the meeting for Science and Literacy Connections, Baltimore, MD.

Gianello, L. (1988). *Getting into gear: Gender inclusive teaching strategies developed by the McClintock Collective.* Canberra: Curriculum Development Centre.

Gobert, J. & Clement, J. (1999). The effects of student-generated diagrams versus student-generated summaries on conceptual understanding of spatial, causal and dynamic knowledge in plate tectonics. *Journal of Research in Science Teaching, 36*, 39-53.

Gunstone, R. (1995). Constructivist learning and the teaching of science. In B. Hand & V. Prain, (Eds.), *Teaching and learning in science: The constructivist classroom.* Sydney: Harcourt Brace.

Hall, G., & Loucks, S. (1979). Teacher concerns as a basis for facilitating and personalizing staff development. In A. Lieberman & L. Miller (Eds.), *Staff development: New demands, new realities, new perspectives* (pp. 245-67). New York: Teachers Press.

Hall, G., & Hord, S. (2001). *Implementing change: Patterns, principles and potholes.* Boston, MA: Allyn & Bacon.

Halliday, M. A. K. (1978). *Language as social semiotic : The social interpretation of language and meaning.* London: Edward Arnold.

Halliday, M. A. K. (1993). Towards a language-based theory of learning. *Linguistics and Education, 5,* 93-116.

Halliday, M. A. K. (1993). Some grammatical problems in scientific English. In M. Halliday & J. Martin, *Writing science: Literacy and discursive power* (pp. 69-75). London: Falmer Press.

Halliday, M. A. K. & Martin, J. R.(1993). *Writing science: Literacy and discursive power.* London: Falmer Press.

Hand, B. (1996). Diagnosis of teacher changes when implementing constructivist teaching/learning approaches. In D. F. Treagust, R. Duit, & B. Fraser (Eds.), *Teaching and learning in science and mathematics* (pp. 212-221). New York: Teachers College Press.

Hand, B., & Craker, G. (1991, July). Language in science classrooms: A base for teachers moving to conceptual change strategics. A paper presented at joint day session National Reading Association Conference and Conference of Australian Science Teachers Association, Adelaide, Australia.

Hand, B., & Prain, V. (1995). Using writing to help improve students' understanding of science knowledge. *School Science Review, 77*(278), 112-117.

Hand, B., Prain, V., Lawrence, C., & Yore, L. D. (1999). A writing in science framework designed to enhance science literacy. *International Journal of Science Education, 21*, 1021-1035.

Hand, B., Prain, V., & Scholes, I. (1997). Light learning. *The Science Teacher,* December, 38-40.

Hand, B., Prain, V., & Wallace, C. S. (2002). Writing tasks influences on extended-response, recall, and higher-level test questions. *Research in Science Education, 32,* 19-34.

Hand, B., Prain,V., & Yore, L. D. (2001). Sequential writing tasks' influence on science learning. In P. Tynjala, L. Mason, & K. Lonka, (Eds.) *Writing as a learning tool* (pp. 105-130). Amsterdam: Kluwer Press.

Hand, B., Wallace, C. S., & Yang, E. M. (in press). Using the Science Writing Heuristic to enhance learning outcomes from laboratory activities in seventh grade science: Quantitative and qualitative aspects. *International Journal of Science Education.*

Hanrahan, M. (1999). Rethinking science literacy: enhancing communication and participation in school science through affirmational dialogue journal writing. *Journal of Research in Science Teaching, 36,* 699-718.

Haraway, D. (1989). *Primate visions.* New York: Routledge.

Haraway, D. (1991). *Simians, cyborgs, and women.* New York: Routledge

Hayes, D. A. (1987). The potential for directing study in combined reading and writing activity. *Journal of Reading Behavior, 19*, 333-352.

Henriques, L. (1997). *A study to define and verify a model of interactive-constructive elementary school teaching.* Unpublished manuscript, University of Iowa.

Hildebrand, G. (1999, April). *Breaking the pedagogical contract: Teachers' and students' voices.* Paper presented at the Annual Meeting of the National Association for Research in Science Teaching, Boston, MA.

Hildebrand, G. (1998). Disrupting hegemonic writing practices in school science: Contesting the right way to write. *Journal of Research in Science Teaching, 35,* 345-362.

Hodson, D. (1998). *Teaching and learning science: Towards a personalized approach.* Buckingham, United Kingdom: Open University Press.

Holliday, W. G., Yore, L. D., & Alvermann, D. E. (1994). The reading-science learning-writing connection: Breakthroughs, barriers and promises. *Journal of Research in Science Teaching, 31,* 877-893.

Jonassen, D., & Carr, C. (2000) Mindtools: Affording multiple knowledge representations for learning. In S. LaJoie (Ed.), *Computers as Cognitive Tools* , Mahwah, NJ: Lawrence Erlbaum Associates. (pp. 165-196).

Kamberelis, G. (1999). Genre development and learning: Children's writing stories, science reports, and poems. *Research in the Teaching of English, 33*, 403-460.

Keller, E., & Longino, H. (Eds.) *Feminism and science*. Oxford, United Kingdom: Oxford University Press.

Kelly, G. J., & Chen, C. (1999). The sound of music: Constructing science as sociocultural practices through oral and written discourse. *Journal of Research in Science Teaching, 36,* 883-915.

Kelly, G. J., Chen, C., & Prothero, W. (2000). The epistemological framing of a discipline: Writing science in university oceanography. *Journal of Research in Science Teaching, 37,* 691-718.

Keys, C. W. (1994). The development of scientific reasoning skills in conjunction with collaborative writing assignments: An interpretive study of six ninth-grade students. *Journal of Research in Science Teaching, 31*, 1003-1022.

Keys, C. W. (1999). Language as an indicator of meaning generation: An analysis of middle school students' written discourse. *Journal of Research in Science Teaching, 36,* 1044-1061.

Keys, C. W. (1999). Revitalizing instruction in scientific genres: Connecting knowledge production with writing to learn science. *Science Education, 83,* 115-130.

Keys, C. W. (2000). Investigating the thinking processes of eighth grade writers during the composition of a scientific laboratory report. *Journal of Research in Science Teaching, 37,* 676-690.

Klein, P. (1999a). Reopening inquiry into cognitive processes in writing-to-learn. *Educational Psychology Review, 11,* 203-270.

Klein, P. (1999b). Learning science through writing: The role of rhetorical structures. The *Alberta Journal of Educational Research, XLV*(2), 132-153.

Klein. P. (2000). Elementary students' strategies for writing-to-learn in science. *Cognition and Instruction,18*(3), 317 - 348

Langer, J. A., & Appleby, A. N. (1987). *How writing shapes thinking: A study of teaching and learning* (Tech. Rep. No. 22). Urbana, IL: National Council of Teachers of English.

Lee, O., & Fradd, S. H. (1998). Science for all, including students from non-English language backgrounds. *Educational Researcher, 27*(3), 1-10.

Lemke, J. (1998). Multiplying meaning: Visual and verbal semiotics in scientific text. In J, Martin & R. Veel (Eds.), *Reading science: Critical and functional perspectives on discourses of science* (pp. 87-113). London: Routledge.

Lemke. J. (2001). Articulating communities: Sociocultural perspectives on science education. *Journal of Research in Science Teaching, 38*, 296-316.

Lemke, J. (2002a). Teaching All the Languages of Science: Words, Symbols, Images, and Actions. http://academic.brooklyn.cuny.edu/education/jlemke/papers/barcelon.htm.

Lemke, J. (2002b) Multimedia genres for science education and scientific literacy. In M. Schleppegrell and C. Colombi, (Eds.), *Developing advanced literacy in first and second languages* (21-44). Hillsdale, NJ: Erlbaum.

Lemke, J. (2002c, September). *Getting critical about science literacies.* Paper presented at the Conference on Ontological, Epistemological, Linguistic and Pedagogical Considerations of Language and Science Literacy: Empowering Research and Informing Instruction, Victoria, British Columbia, Canada.

Levins, L., Pegg., J., & Creedy. L. (1994). A program for professional development of secondary science teachers. *Australian Science Teachers Journal, 40*(1), 37-41.

Lincoln, Y. & Guba, E.(1985). *Naturalistic inquiry*. Newbury Park: Sage Publications.

Linn, M. (2000) Designing the knowledge integration environment, *International Journal of Science Education* 22, pp. 781-796.

Loucks, S. (1996). Professional development for science education: A critical and immediate challenge. In R. Bybee (Ed.), *National standards and the science curriculum* (pp. 224-227). Dubuque, Iowa: Kendall/Hunt Publishing.

Loucks-Horsley, S., Hewson, P., Love, N., & Stiles, K. (1998*). Designing professional development for teachers of science and mathematics.* Thousand Oaks, CA: Corwin Press.

Martin, J. (1999). Mentoring semogenesis: 'genre-based' literacy pedagogy. In F. Christie (Ed.) *Pedagogy and the shaping of consciousness: Linguistic and social processes*. (Open Linguistics Series, pp. 123-155). London: Cassell.

Martin, J. & Veel, R. (1998). *Reading science: critical and functional perspectives on discourses of science.* London: Routledge.

Martin, N. (1992) Language across the curriculum: Where it began and what it promises. In A. Herrington & C. Moran (Eds.), *Writing, teaching, and learning in the disciplines* (pp. 6-21). New York: Modern Language Association of America.

Mason, L., & Buscolo, P. (2000). Writing and conceptual change: What changes? *Instructional Science, 28,* 199-226.

McCrindle, A. R. & Christensen, C. A. (1995). The impact of learning journal on metacognitive and cognitive processes and learning performance. *Learning and Instruction, 5,* 167-185.

MacDonald, B., Atkin, R., Jenkins, D. & Kemmis. S. (1997). Computer Assisted learning: its educational potential. In R. Hopper (Ed.), *The national development program in computer assisted learning.* London: Council for Educational Technology.

McDonald, H. R. & Purdy, R. A. (1989). Description and evaluation of a short writing assignment in historical geology. *Journal of Geological Education, 37,* 117-120.

Merriam, S. B. (1988). *Case study research in education: A qualitative approach.* San Francisco, CA: Jossey-Bass.

Michaels, S., & Sohmer, R. (2000). Narratives and inscriptions: cultural tools, power and powerful sense-making. In B. Cope & M. Kalantzis (Eds.). *Multiliteracies: literacy learning and the design of social futures* (pp. 267-288). London: Routledge.

Miles, M. B. & Huberman, A. M. (1994). *Qualitative data analysis: A sourcebook of new methods* (2nd ed.). Newbury Park, CA. Sage Publications.

Millar, R. (1998). Rhetoric and reality: What practical work in science education is *really* for. In J. Wellington (ed.) *Practical work in science: Which way now?* New York: Rutledge.

Minstrell, J. (1989). Teaching science for understanding. In L. B. Resnick & L.E. Klopfer (Eds.), *Towards the thinking curriculum: Current cognitive research* (1989 Yearbook; pp. 129-149). Alexandria, VA: Association for Supervision and Curriculum Development.

Myers, G. (1990). *Writing biology: Texts in the social construction of biology knowledge.* Madison: University of Wisconsin Press.

National Research Council. (1996). *National Science Education Standards.* Washington, D.C.: National Academy Press.

Norris, S. P., & Phillips, L. M. (2001, March). *How literacy in its fundamental sense is central to scientific literacy.* Paper presented at the annual meeting for the National Association for Research in Science Teaching, St. Louis, MI.

Novak, J. (1998). *Learning, creating and using knowledge: Concept map tools in schools and corporations.* Hillsdale, NJ: Erlbaum.

Osborne, R., & Wittrock, M. (1983). Learning science: A generative process. *Science Education, 67,* 489-508.

Ogborn, J., Kress, G., Martins, I., & McGillicuddy, K. (1996). *Explaining science in the classroom.* Buckingham, United Kingdom: Open University Press.

Parker, L. (1992). Language in science education: implications for teachers. *Australian Science Teachers Journal, 38* (2), 29-32.

Prain, V. (1995). Writing for learning in secondary science. In B. Hand & V. Prain, (Eds.), *Teaching and learning in science: The constructivist classroom* (pp. 63-81). Harcourt Brace, Sydney.

Prain, V., & Hand, B. (1996a). Writing and learning in secondary science: Rethinking practices. *Teaching and Teacher Education, 12,* 609-626.

Prain, V., & Hand, B. (1996b). Writing for learning in the junior secondary science classroom: Issues arising from a case study. *International Journal of Science Education, 18,* 117-128.

Prain, V., & Hand, B. (1999). Students perceptions of writing for learning in secondary school science. *Science Education, 83,* 151-162.

Prain, V., Hand, B., & Kay, S. (1996). Writing for learning in physics. *The Physics Teacher 35*(1), 40-41.

Resnik, L. B., & Klopfer, L.E. (1989). *Towards the thinking curriculum: Current cognitive research* (1989 Yearbook). Alexandria, VA: Association for Supervision of Curriculum Development.

Rivard, L. P. (1994). A review of writing to learn in science: Implications for practice and research. *Journal of Research in Science Teaching, 31,* 969-983.

Rivard, L. P., & Straw, S. B. (2000). The effect of talk and writing on science learning: An exploratory study. *Science Education, 84,* 566-593.

Roth, K. (1990). Developing meaningful conceptual understanding in science. In B.F. Jones & L. Idol (Eds.), *Dimensions of thinking and cognitive instruction* (pp. 139-175). Hillsdale, NJ: Erlbaum.

Roth, W-M. (1995). *Authentic school science* (Vol. 1). Dordrecht, The Netherlands: Kluwer Academic Publishers.

Rowell, P.A. (1997). Learning in school science: The promises and practices of writing. *Studies in Science Education, 30,* 19-56.

Russell, T. & McGuigan, L. (2001). Promoting understanding through representational redescription: an illustration referring to young pupils' ideas about gravity. In D. Psillos, P. Kariotoglou, V. Tselfes, G. Bisdikian, G. Fassoulopoulos, E. Hatzikraniotis, E. Kallery (Eds.), *Science Education Research in the Knowledge-Based Society. Proceedings of the Third International Conference of the ESERA* (pp. 600-602). Thessaloniki, Greece: Aristotle University.

Santa, C. M. & Havens, L. T. (1991). Learning through writing. In Santa, C. M. & Alvermann, D. E. (Eds.), Science learning: Processes and applications. Newark, DE: International Reading Association.

Scardamalia, M., & Bereiter, C. (1986). Helping students become better writers. *School Administrator, 42*(4), 16 & 26.

Scheppegrell, M. (1998). Grammar as resource: Writing as description. *Research in the Research in the Teaching of English, 25*, 67-96.

Schumacher, G., & Nash, J. (1991). Conceptualizing and measuring knowledge change due to writing. *Research in the Teaching of English, 25,* 67-96.

Shepherd, L. (1993). *Lifting the veil: The feminine face of science.* Boston: Shambhala Publications.

Shepardson, D. P., & Britsch, S. J. (2001). The role of children's journals in elementary school science activities. *Journal of Research in Science Teaching, 38*, 43-69.

Shymansky, J., & Kyle, W. (1992). Establishing a research agenda: Critical issues of science curriculum reform. *Journal of Research in Science Teaching, 29*, 749-778.

Stadler, H., Benke, G., & Duit, R. (2001). How do boys and girls use language in physics classes? In H. Behrendt, H. Dahncke, R. Duit, W. Graber, M. Komorek, A. Kross, P. Reiska, (Eds.) *Research in science education – past, present, and future* (pp. 283-288). Amsterdam: Kluwer Press.

Stencel, J. E. (1998). An interactive lecture notebook - twelve ways to improve students' grades. *Journal of College Science Teaching, 27,* 343-345.

Stenning, K. (1998). Representation and conceptualisation in educational communication. In M. W. van Someren & P. Reimann & H. P. A. Boshuizen & T. de Jong (Eds.), *Learning with multiple representations* (pp. 320-333). Amsterdam: Pergamon.

Stronck, D. R., & Koller, G. R. (1991). Evaluating the effectiveness of an inservice education program through the use of materials. *Journal of Research in Science Teaching, 18,* 403-408.

Stylianidou, F., Boohan, R., & Ogborn, J. (2001). *Computer modelling and simulation in science lessons: Using research into teachers' transformations to inform training.* In D. Psillos, P. Kariotoglou, V. Tselfes, G. Bisdikian, G. Fassoulopoulos, E. Hatzikraniotis, E. Kallery (Eds.), Science education research in the knowledge-based society (pp. 159-161). Proceedings of the Third International Conference of the European Science Education Research Association, Aristotle University, Thessonaliki, Greece.

Sutton, C. (1992). *Words, science and learning.* Buckingham, United Kingdom: Open University Press.

Sutton, C. (1996). Beliefs about science and beliefs about language. *International Journal of Science Education, 18*, 1-18.

Tsoi, M. Y., & Keys, C. W. (2001, March) *Exploring tenth grade students' use of the Science Writing Heuristic student template for thinking about laboratory activities.* Paper presented at the annual meeting of the National Association for Research in Science Teaching, St. Louis, MO.

Tucknott, J. M. & Yore, L. D. (1999, March). *The effects of writing activities on grade 4 children's understanding of simple machines, inventions, and inventors.* Paper presented at the annual meeting for the National Association of Research in Science Teaching, Boston, MA.

Tynjala, P., Mason, L. & Lonka, K. (2001). *Writing as a learning tool.* Amsterdam: Kluwer Press.

Tytler, R., & Waldrip, B. (2002). Schools' experience of change. *Investigating, 18*(4), 23-26.

Varelas, M., Becker, J. Luster, B, & Wenzel, S. (2002). When genres meet: Inquiry into a sixth-grade urban science class. *Journal of Research in Science Teaching, 39,* 579-605.

Varelas, M., Pappas, C. C., Barry, A., & Rife, A. (2002, April). *Young urban children's science texts and pictures: Emergent understandings and semiotic representations.* Paper presented at the annual meeting for the National Association of Research in Science Teaching, New Orleans, LA.

Veel, R. (1996). Learning how to mean — scientifically speaking. In F. Christie (Ed.) *Genre and institutions: The language of work and schooling*. London: Cassell.

Unsworth, L. (1997). Explaining explanations: enhancing science learning and literacy development. *Australian Science Teachers Journal, 43* (1), 34-49.

Unsworth, L. (1999) Explaining school science in book and CD ROM formats: Using semiotic analyses to compare the textual construction of knowledge. *International Journal of Instructional Media, 26*(2),159-179.

Unsworth, L. (2000). Investigating subject-specific literacies in school learning. In L. Unsworth, (Ed.). *Researching language in schools and communities.* London: Cassell.

Unsworth, L. (2001). *Teaching multiliteracies across the curriculum: Changing contexts of text and image in classroom practice*. Buckingham, United Kingdom: Open University Press.

von Glasersfeld,E.(1988). The reluctance to change a way of thinking. *The Irish Journal of Psychology, 9*(1), 83-90.

von Glasersfeld, E. (1989) Cognition, construction of knowledge and teaching. *Synthese, 80,* 121-140.

Vygotsky, L. (1962). *Thought and language*, (E. Hanfmann & G. Vakar, Trans.). Cambridge, MA: MIT Press.

Vygotsky, L. S. (1978). Mind in society: The development of higher psychological processes. Cole, M., John-Steiner, V., Scribner, S., & Souberman, E. (Eds.) Cambridge, MA: Harvard University Press.

Walvoord, B. (1982). *Helping students write well: A guide for teachers in all disciplines*. New York: Modern Language Association in America.

Wallace, C. S. (in press). An illumination of the roles of hands-on activities, discussion, text reading, and writing in constructing biology knowledge in seventh grade. *School Science and Mathematics.*

Wellington, J. (Ed.). (1998). *Practical work in school science: Which way now?* London: Routeledge.

Wellington, J. & Osborne, J. (2001). *Language and literacy in science education.* Buckingham, United Kingdom: Open University Press.

Yore, L. D. (2001). What is meant by constructivist science teaching and will the science education community stay the course for meaningful reform? Electronic Journal of Science Education, 5(4), Online journal:http://unr.edu/homepage/crowther/ejse.

SUBJECT INDEX

Science & Technology Education Library

Series editor: William W. Cobern, *Western Michigan University, Kalamazoo, U.S.A.*

Publications

1. W.-M. Roth: *Authentic School Science*. Knowing and Learning in Open-Inquiry Science Laboratories. 1995 ISBN 0-7923-3088-9; Pb: 0-7923-3307-1
2. L.H. Parker, L.J. Rennie and B.J. Fraser (eds.): *Gender, Science and Mathematics*. Shortening the Shadow. 1996 ISBN 0-7923-3535-X; Pb: 0-7923-3582-1
3. W.-M. Roth: *Designing Communities*. 1997 ISBN 0-7923-4703-X; Pb: 0-7923-4704-8
4. W.W. Cobern (ed.): *Socio-Cultural Perspectives on Science Education*. An International Dialogue. 1998 ISBN 0-7923-4987-3; Pb: 0-7923-4988-1
5. W.F. McComas (ed.): *The Nature of Science in Science Education*. Rationales and Strategies. 1998 ISBN 0-7923-5080-4
6. J. Gess-Newsome and N.C. Lederman (eds.): *Examining Pedagogical Content Knowledge*. The Construct and its Implications for Science Education. 1999 ISBN 0-7923-5903-8
7. J. Wallace and W. Louden: *Teacher's Learning*. Stories of Science Education. 2000 ISBN 0-7923-6259-4; Pb: 0-7923-6260-8
8. D. Shorrocks-Taylor and E.W. Jenkins (eds.): *Learning from Others*. International Comparisons in Education. 2000 ISBN 0-7923-6343-4
9. W.W. Cobern: *Everyday Thoughts about Nature*. A Worldview Investigation of Important Concepts Students Use to Make Sense of Nature with Specific Attention to Science. 2000 ISBN 0-7923-6344-2; Pb: 0-7923-6345-0
10. S.K. Abell (ed.): *Science Teacher Education*. An International Perspective. 2000 ISBN 0-7923-6455-4
11. K.M. Fisher, J.H. Wandersee and D.E. Moody: *Mapping Biology Knowledge*. 2000 ISBN 0-7923-6575-5
12. B. Bell and B. Cowie: *Formative Assessment and Science Education*. 2001 ISBN 0-7923-6768-5; Pb: 0-7923-6769-3
13. D.R. Lavoie and W.-M. Roth (eds.): *Models of Science Teacher Preparation*. Theory into Practice. 2001 ISBN 0-7923-7129-1
14. S.M. Stocklmayer, M.M. Gore and C. Bryant (eds.): *Science Communication in Theory and Practice*. 2001 ISBN 1-4020-0130-4; Pb: 1-4020-0131-2
15. V.J. Mayer (ed.): *Global Science Literacy*. 2002 ISBN 1-4020-0514-8
16. D. Psillos and H. Niedderer (eds.): *Teaching and Learning in the Science Laboratory*. 2002 ISBN 1-4020-1018-4
17. J.K. Gilbert, O. De Jong, R. Justi, D.F. Treagust and J.H. Van Driel (eds.): *Chemical Education: Towards Research-based Practice*. 2003 ISBN 1-4020-1112-1
18. A.E. Lawson: *The Neurological Basis of Learning, Development and Discovery*. Implications for Science and Mathematics Instruction. 2003 ISBN 1-4020-1180-6
19. D.L. Zeidler (ed.): *The Role of Moral Reasoning on Socioscientific Issues and Discourse in Scientific Education*. 2003 ISBN 1-4020-1411-2

Science & Technology Education Library

Series editor: William W. Cobern, *Western Michigan University, Kalamazoo, U.S.A.*

20. P.J. Fensham: *Defining an Identity. The Evolution of Science Education as a Field of Research.* 2003 ISBN 1-4020-1467-8
21. D. Geelan: *Weaving Narrative Nets to Capture Classrooms.* Multimethod Qualitative Approaches for Educational Research. 2003 ISBN 1-4020-1776-6; Pb: 1-4020-1468-7
22. A. Zohar: *Higher Order Thinking in Science Classrooms: Students' Learning and Teachers' Professional Development.* 2004 ISBN 1-4020-1852-5; Pb: 1-4020-1853-3

KLUWER ACADEMIC PUBLISHERS – DORDRECHT / BOSTON / LONDON